A. M. Yaglom and I. M. Yaglom

CHALLENGING MATHEMATICAL PROBLEMS WITH ELEMENTARY SOLUTIONS

Volume I

Combinatorial Analysis and Probability Theory

Translated by James McCawley, Jr.

Revised and edited by Basil Gordon

DOVER PUBLICATIONS, INC.
NEW YORK

Published in Canada by General Publishing Company,
Ltd., 30 Lesmill Road, Don Mills, Toronto, Ontario.

Published in the United Kingdom by Constable and Company, Ltd., 10 Orange Street, London WC2H 7EG.

This Dover edition, first published in 1987, is an un-
abridged and unaltered republication of the edition pub-
lished by Holden-Day, Inc., San Francisco, in 1964. It was
published then as part of the *Survey of Recent East
European Mathematical Literature,* a project conducted by
Alfred L. Putnam and Izaak Wirszup, Dept. of Mathematics,
The University of Chicago, under a grant from The National
Science Foundation. It is reprinted by special arrangement
with Holden-Day, Inc., 4432 Telegraph Ave., Oakland,
California 94609.

Originally published as *Neelementarnye Zadachi v Elementarnom Izlozhenii* by the Government Printing House
for Technical-Theoretical Literature, Moscow, 1954.

Manufactured in the United States of America
Dover Publications, Inc., 31 East 2nd Street, Mineola,
N.Y. 11501

Library of Congress Cataloging-in-Publication Data

Yaglom, A. M.
    [Neelementarnye zadachi v elementarnom izlozhenii.
English]
    Challenging mathematical problems with elementary
solutions / A. M. Yaglom and I. M. Yaglom : translated by
James McCawley, Jr. : revised and edited by Basil Gordon.
        p.   cm.
    Translation of: Neelementarnye zadachi v elementar-
nom izlozhenii. Reprint. Originally: San Francisco :
Holden-Day, 1964–1967.
    Bibliography: p.
    Includes indexes.
    Contents: v. 1. Combinatorial analysis and probability
theory—v. 2. Problems from various branches of mathematics.
    ISBN 0-486-65536-9 (pbk. : v. 1). ISBN 0-486-65537-7
(pbk. : v. 2)
    1. Combinatorial analysis—Problems, exercises, etc.
2. Probabilities—Problems, exercises, etc.   3. Mathematics—Problems, exercises, etc.   I. Yaglom, I. M.
(Isaac Moiseevich), 1921–   II. Gordon, Basil.
III. Title.
QA 164.I1613   1987
511'.6—dc19   133562   87-27298
                                        CIP

PREFACE TO THE AMERICAN EDITION

This book is the first of a two-volume translation and adaptation of a well-known Russian problem book entitled *Non-Elementary Problems in an Elementary Exposition.** The first part of the original, *Problems on Combinatorial Analysis and Probability Theory*, appears as Volume I, and the second part, *Problems from Various Branches of Mathematics*, as Volume II. The authors, Akiva and Isaak Yaglom, are twin brothers, prominent both as mathematicians and as expositors, whose many excellent books have been exercising considerable influence on mathematics education in the Soviet Union.

This adaptation is designed for mathematics enthusiasts in the upper grades of high school and the early years of college, for mathematics instructors or teachers and for students in teachers' colleges, and for all lovers of the discipline; it can also be used in problem seminars and mathematics clubs. Some of the problems in the book were originally discussed in sections of the School Mathematics Circle (for secondary school students) at Moscow State University; others were given at Moscow Mathematical Olympiads, the mass problem-solving contests held annually for mathematically gifted secondary school students.

The chief aim of the book is to acquaint the reader with a variety of new mathematical facts, ideas, and methods. The form of a problem book has been chosen to stimulate active, creative work on the materials presented.

The first volume contains 100 problems and detailed solutions to them. Although the problems differ greatly in formulation and method of solution, they all deal with a single branch of mathematics: combinatorial analysis. While little or no work on this subject is done in American high schools, no knowledge of mathematics beyond what is imparted in a good high school course is required for this book. The authors have tried to outline the elementary methods of combinatorial analysis with some completeness, however. Occasionally, when needed, additional explanation is given before the statement of a problem.

* *Neelementarnye zadachi v elementarnom izlozhenii*, Moscow: Gostekhizdat, 1954.

v

Thus the majority of the problems in this book and in its companion volume represent questions in higher ("non-elementary") mathematics that can be solved with elementary mathematics. Most of the problems in this volume are not too difficult and resemble problems encountered in high school. The last three sections, however, contain some very difficult problems. Before going on to the problems, the reader should consult the "Suggestions for Using the Book."

The book was translated by Professor James McCawley, Jr., of the University of Chicago and edited and revised by Professor Basil Gordon of the University of California at Los Angeles.

Problem 85 was sent by the Russian authors for inclusion in the American edition, and appears here for the first time. A number of revisions have been made by the editor:

1. In order to make volume I self-contained, some problems were transferred to volume II. To replace these, problems 1, 3, 12, and 100 were added. Problem 12, in which the principle of inclusion and exclusion is presented, is intended to unify the treatment of several subsequent problems.

2. Some of the problems have been restated in order to illustrate the same ideas with smaller numbers.

3. The introductory remarks to section 1, 2, 6, and 8 have been rewritten so as to explain certain points with which American readers might not be familiar.

4. Adaptation of this book for American use has involved these customary changes: References to Russian money, sports, and so forth have been converted to their American equivalents; some changes in notation have been made, such as the introduction of the notation of set theory where appropriate; some comments dealing with personalities have been deleted; and Russian bibliographical references have been replaced by references to books in English, whenever possible.

The editor wishes to thank Professor E. G. Straus for his helpful suggestions made during the revision of the book. The Survey wishes to express its particular gratitude to Professor Gordon for the valuable improvements he has introduced.

SUGGESTIONS
FOR USING THE BOOK

This book contains one hundred problems. The statements of the problems are given first, followed by a section giving complete solutions. Answers and hints are given at the end of the book. For most of the problems the reader is advised to find a solution by himself. After solving the problem, he should check his answer against the one given in the book. If the answers do not coincide, he should try to find his error; if they do, he should compare his solution with the one given in the solutions section. If he does not succeed in solving the problem alone, he should consult the hints in the back of the book (or the answer, which may also help him to arrive at a correct solution). If this is still no help, he should turn to the solution. It should be emphasized that an attempt at solving the problem is of great value even if it is unsuccessful: it helps the reader to penetrate to the essence of the problem and its difficulties, and thus to understand and to appreciate better the solution presented in the book.

But this is not the best way to proceed in all cases. The book contains many difficult problems, which are marked, according to their difficulty, by one, two, or three asterisks. Problems marked with two or three asterisks are often noteworthy achievements of outstanding mathematicians, and the reader can scarcely be expected to find their solutions entirely on his own. It is advisable, therefore, to turn straight to the hints in the case of the harder problems; even with their help a solution will, as a rule, present considerable difficulties.

The book can be regarded not only as a problem book, but also as a collection of mathematical propositions, on the whole more complex than those assembled in Hugo Steinhaus's excellent book, *Mathematical Snapshots* (New York: Oxford University Press, 1960), and presented in the form of problems together with detailed solutions. If the book is used in this way, the solution to a problem may be read after its statement is clearly understood. Some parts of the book, in fact, are so written that this is the best way to approach them. Such, for example, are problems 53 and 54, problems 83 and 84, and, in general, all problems marked with three asterisks. Sections VII and VIII could also be treated in this way.

The problems are most naturally solved in the order in which they occur. But the reader can safely omit a section he does not find interesting. There is, of course, no need to solve all the problems in one section before passing to the next.

This book can well be used as a text for a school or undergraduate mathematics club studying combinatorial analysis and its applications to probability theory. In this case the additional literature cited in the text will be of value. While the easier problems could be solved by the partic-pants alone, the harder ones should be regarded as "theory." Their solutions might be studied from the book and expounded at the meetings of the club.

INDEX OF PROBLEMS GIVEN
IN THE MOSCOW MATHEMATICAL OLYMPIADS

The Olympiads are conducted in two rounds: the first is an elimination round, and the second is the core of the competition.

Olympiads	Round I	Round II	Olympiads	Round I	Round II
For 7th and 8th graders			For 9th and 10th graders		
VI (1940)	—	16, 35a	I (1935)	—	6, 27
VIII (1945)	—	62a	II (1936)	—	17
X (1947)	20	—	III (1937)	—	47
XIII (1950)	—	54[1]	IV (1938)	2	13a, 45a
			V (1939)	—	45b[2]
			VI (1940)	4	15
			VIII (1945)	—	62b
			X (1947)	49a	—
			XI (1948)	—	26
			XII (1949)	—	91a

[1] For $n = 10$.
[2] For $n = 5$.

PROBLEMS

CHALLENGING MATHEMATICAL PROBLEMS WITH ELEMENTARY SOLUTIONS

Volume I

Combinatorial Analysis and Probability Theory

PROBLEMS

The problems in this volume are related by the fact that in nearly all of them we are required to answer a question of "how many?" or "in how many ways?" Such problems are called *combinatorial*, as they are exercises in calculating the number of different combinations of various objects. The branch of mathematics which deals with such problems is called *combinatorial analysis*.

In the solutions to many of the problems, the following notation is used. Let n and k be integers such that $0 \leq k \leq n$. Put

$$\binom{n}{k} = \frac{n!}{k!\,(n-k)!} = \frac{n(n-1)(n-2)\cdots(n-k+1)}{k!}$$

The symbol $\binom{n}{k}$ may be read as "the binomial coefficient n over k". (Indeed these numbers occur as coefficients in the binomial theorem; in section V we will study them from that point of view.) For example, $\binom{7}{3} = \frac{7\cdot6\cdot5}{3\cdot2\cdot1} = 35$. By virtue of the convention that $0! = 1$, we have $\binom{n}{0} = \frac{n!}{0!\,n!} = 1$, and similarly $\binom{n}{n} = 1$. In general $\binom{n}{k} = \binom{n}{n-k}$, as is easily seen from the definition.

We wish to point out here that $\binom{n}{k}$ is the number of ways in which k objects can be selected from a given set of n objects. To begin with a concrete example, suppose we have a set S of five elements, say $S = \{a,b,c,d,e\}$, and we wish to select a subset T of two elements from S (thus $n = 5$, $k = 2$). We can easily list all such sets T; they are $\{a,b\}$, $\{a,c\}$, $\{a,d\}$, $\{a,e\}$, $\{b,c\}$, $\{b,d\}$, $\{b,e\}$, $\{c,d\}$, $\{c,e\}$, and $\{d,e\}$. Thus there are altogether 10 possibilities for the selection, and indeed $\binom{5}{2} = \frac{5\cdot4}{2\cdot1} = 10$.

In the general case, where S has n elements and T has k elements, let us introduce the notation $C_n^{\,k}$ for the number of such sets T. Thus

our object will be to prove that $C_n^{\ k} = \binom{n}{k}$. For this purpose it is convenient to introduce the notion of an *ordered set*, i.e., a set whose elements
are written down in a definite order. Two ordered sets are said to be
equal if and only if they consist of the same elements in the same order.
Thus in the above example, the sets $\{a,b\}$ and $\{b,a\}$ are the same, but
considered as ordered sets they are different. Now let $T = \{a_1, a_2, \ldots, a_k\}$
be a set of k elements, and let us calculate the number of ways in which
these elements can be ordered. There are k possibilities for the first
element. Once it has been chosen, there are $k - 1$ possibilities for the
second element; once the first two elements have been chosen, there are
$k - 2$ possibilities for the third element, etc. Hence there are altogether
$k(k - 1)(k - 2) \cdots 2.1 = k!$ orderings. From this it follows that if $P_n^{\ k}$
is the number of *ordered k-element subsets of S*, then

(1) $$P_n^{\ k} = k! \, C_n^{\ k}.$$

But we can calculate $P_n^{\ k}$ directly by reasoning similar to the above. The
first element of the ordered set T can be chosen from S in n ways. Once it
is chosen, the second element can be chosen in $n - 1$ ways, etc. Hence

$$P_n^{\ k} = n(n - 1)(n - 2) \cdots (n - k + 1),$$

where there are k factors on the right. From (1) we now obtain

$$C_n^{\ k} = \frac{P_n^{\ k}}{k!} = \frac{n(n - 1) \cdots (n - k + 1)}{k!} = \binom{n}{k},$$

completing the proof. Note that when $k = 0$, we are allowing T to be the
empty set; thus the fact that $\binom{n}{0} = 1$ is not a paradox.

Because of the above result, the quantity $\binom{n}{k} = C_n^{\ k}$ is often called
the number of combinations of n objects taken k at a time. Similarly, $P_n^{\ k}$
is referred to as the number of permutations of n objects taken k at a time
(the word *permutation* being an older term for ordered set).

I. INTRODUCTORY PROBLEMS

1. Three points in the plane are given, not all on the same straight line.
How many lines can be drawn which are equidistant from these points?

2. Four points in space are given, not all in the same plane. How many
planes can be drawn which are equidistant from these points?

3. Four points in the plane are given, not all on the same straight line, and not all on a circle. How many straight lines and circles can be drawn which are equidistant from these points? (By the distance from a point P to a circle c with center O we mean the length of the segment PQ, where Q is the point where the ray from O in the direction OP meets c.

4. Five points in space are given, not all in the same plane, and not all on the surface of a sphere. How many planes and spheres can be drawn which are equidistant from these points? (By the distance from a point P to a sphere Σ with center O, we mean the length of the segment PQ, where Q is the point where the ray from O in the direction OP meets Σ.)

5. How many spheres are tangent to the planes of all the faces of a given tetrahedron T?

6. Six colors of paint are available. Each face of a cube is to be painted a different color. In how many different ways can this be done if two colorings are considered the same when one can be obtained from the other by rotating the cube?

7. In how many different ways can 33 boys be divided into 3 football teams of 11 boys each?

8. A store sells 11 different flavors of ice cream. In how many ways can a customer choose 6 ice cream cones, not necessarily of different flavors?

9. A group of 11 scientists are working on a secret project, the materials of which are kept in a safe. They want to be able to open the safe only when a majority of the group is present. Therefore the safe is provided with a number of different locks, and each scientist is given the keys to certain of these locks. How many locks are required, and how many keys must each scientist have?

10. The integers from 1 to 1000 are written in order around a circle. Starting at 1, every fifteenth number is marked (that is, 1, 16, 31, etc.). This process is continued until a number is reached which has already been marked. How many unmarked numbers remain?

11a. Among the integers from 1 to 10,000,000,000 which are there more of: those in which the digit 1 occurs or those in which it does not occur?

    b. If the integers from 1 to 222,222,222 are written down in succession, how many 0's are written?

II. THE REPRESENTATION OF INTEGERS AS SUMS AND PRODUCTS

In solving some of the problems of this section, the following notation will prove useful.

The symbol [x] (read "the integral part of x") denotes the greatest integer which is $\leqq x$. Thus, for example,

$$[\tfrac{3}{2}] = 1, \ [10.85] = 10, \ [5] = 5, \ [-8.2] = -9, \text{ etc.}$$

The symbol $N(x)$ (read "nearest integer to x") denotes the integer closest to x. Thus, for example, $N(5.4) = 5$, $N(8.73) = 9$, $N(6) = 6$, $N(-2.8) = -3$.

It is clear that $N(x)$ is equal to $[x]$ or $[x] + 1$ according as $x - [x]$ is less than or greater than $\tfrac{1}{2}$. In the case when $x - [x] = \tfrac{1}{2}$, $N(x)$ could be taken to mean either $[x]$ or $[x] + 1$; in this book we will make the convention that $N(x) = [x] + 1$ for such values of x. It can then be verified that $N(x) = [2x] - [x]$.

If A and B are two sets, we denote by $A \cup B$ (read "*A* union *B*" or "*A* cup *B*") the set of all elements in A or B (or both). We call $A \cup B$ the *union* or *sum* of A and B. In fig. 1, where A and B are represented by two

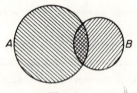

Fig. 1

discs, $A \cup B$ is the entire shaded region. By $A \cap B$ (read "*A* intersect *B*" or "*A* cap *B*") we mean the set of all elements that are in both A and B. In fig. 1 the set $A \cap B$, which is called the *intersection* or *product* of A and B, is the doubly shaded region.

More generally, if $A_1, \ldots, A_m$ are sets, we denote by $A_1 \cup \cdots \cup A_m$ the set of all elements in at least one of the sets $A_1, \ldots, A_m$. By $A_1 \cap \cdots \cap A_m$ we mean the set of all elements which are in all the sets $A_1, \ldots, A_m$. We call $A_1 \cup \cdots \cup A_m$ the union, and $A_1 \cap \cdots \cap A_m$ the intersection, of $A_1, \ldots, A_m$.

12a. For any finite set S, let $\#(S)$ denote the number of elements of S (read "order of S" or "cardinality of S"). Prove that if A and B are finite sets, then

$$\#(A \cup B) = \#(A) + \#(B) - \#(A \cap B).$$

b. Prove that if A, B, and C are finite sets, then

$$\#(A \cup B \cup C) = \#(A) + \#(B) + \#(C) - \#(A \cap B)$$
$$- \#(A \cap C) - \#(B \cap C) + \#(A \cap B \cap C).$$

c.* Prove that if $A_1, A_2, \ldots, A_m$ are finite sets, then

$$\#(A_1 \cup A_2 \cup \cdots \cup A_m) = \#(A_1) + \#(A_2) + \cdots + \#(A_m)$$
$$- \#(A_1 \cap A_2) - \#(A_1 \cap A_3) - \cdots$$
$$- \#(A_{m-1} \cap A_m) + \#(A_1 \cap A_2 \cap A_3)$$
$$+ \#(A_1 \cap A_2 \cap A_4) + \cdots$$
$$+ (-1)^{m-1} \#(A_1 \cap A_2 \cdots \cap A_m).$$

The right-hand side of this formula is formed in the following way. First we have the terms $\#(A_i)$, where $1 \leq i \leq m$. Then we have the terms $- \#(A_i \cap A_j)$, where $1 \leq i < j \leq m$ (there are $\binom{m}{2}$ such terms, since there are $\binom{m}{2}$ ways of selecting the two integers i, j from the numbers $1, \ldots, m$.) Then we have the terms $\#(A_i \cap A_j \cap A_k)$, where $1 \leq i < j < k \leq m$ (there are $\binom{m}{3}$ of these). Next come the terms $- \#(A_i \cap A_j \cap A_k \cap A_l)$, where $1 \leq i < j < k < l \leq m$. We proceed in this way until finally the expression comes to an end when we reach the term $(-1)^{m-1} \#(A_1 \cap A_2 \cap \cdots \cap A_m)$. Part a above is the case $m = 2$, and part b is the case $m = 3$.

This formula is often called the *principle of inclusion and exclusion*.

13a. How many positive integers less than 1000 are divisible neither by 5 nor by 7?

b. How many of these numbers are divisible neither by 3 nor by 5 nor by 7?

14.* How many positive integers ≤ 1260 are relatively prime to 1260?

15. How many positive integers $x \leq 10,000$ are such that the difference $2^x - x^2$ is not divisible by 7?

16. How many different pairs of integers x, y between 1 and 1000 are such that $x^2 + y^2$ is divisible by 49? Here the pairs (x,y) and (y,x) are not to be considered different.

17.* In how many ways can the number 1,000,000 be expressed as a product of three positive integers? Factorizations which differ only in the order of the factors are not to be considered different.

18.* How many divisors does the number 18,000 have (including 1 and 18,000 itself)? Find the sum of all these divisors.

---

* See explanation of asterisks on page vii.

19. How many pairs of positive integers A, B are there whose least common multiple is 126,000? Here (A,B) is to be considered the same as (B,A).

20. Find the coefficients of x^{17} and x^{18} in the expansion of $(1 + x^5 + x^7)^{20}$.

21. In how many ways can a quarter be changed into dimes, nickels, and pennies?

In problems 22–32, the letter n always denotes a positive integer.

22. In how many ways can n cents be put together out of pennies and nickels?

23.** In how many ways can a total postage of n cents be put together using

    a. 1-, 2-, and 3-cent stamps?

    b. 1-, 2-, and 5-cent stamps?

24.** In how many ways can a 100-dollar bill be changed into 1-, 2-, 5-, 10-, 20-, and 50-dollar bills?

25. In how many ways can a number n be represented as a sum of two positive integers if representations which differ only in the order of the terms are considered to be the same?

26. How many solutions in integers does the inequality

$$|x| + |y| < 100$$

have? Here the solutions (x,y) and (y,x) are to be considered different when $x \neq y$.

27. In how many ways can the number n be written as a sum of three positive integers if representations differing in the order of the terms are considered to be different?

28a. In how many ways can the number n be represented as a sum of 3 nonnegative integers x, y, z, if representations differing only in the order of the terms are *not* considered different?

    b. How many such representations are there if x, y, and z are required to be positive?

29.* How many positive integral solutions of the equation $x + y + z = n$ satisfy the inequalities $x \leq y + z, y \leq x + z, z \leq x + y$? Here solutions differing only in the order of the terms are to be considered as different.

30.** How many incongruent triangles are there with perimeter n if the lengths of the sides are integers?

31a.* How many different solutions in positive integers does the equation

$$x_1 + x_2 + x_3 + \cdots + x_m = n$$

have?

b. How many solutions in nonnegative integers does the equation

$$x_1 + x_2 + x_3 + \cdots + x_m = n$$

have?

Remark. Problem 27 is a special case of problem 31a (that corresponding to $m = 3$).

To conclude this set of problems we will present four general theorems dealing with the representation of numbers as sums of positive integers. The first three of them are by Leonhard Euler (1707–1783), one of the greatest mathematicians of the Eighteenth Century, who derived a great many important results in the most diverse branches of mathematics.[1] A series of similar theorems is contained in chapter XVI of Euler's book *Introductio in Analysin Infinitorum*. Euler proves his theorems by the use of an interesting general method (the "method of generating functions"); these proofs are different from the more elementary ones presented in this book as solutions to problems 32 and 33.

In problems 32 and 33 representations of a number n as a sum which differ only in the order of the terms are considered to be the same. Such representations are called *partitions* of n, and the terms are called *parts*.

32a.* Prove that the number of partitions of n into at most m parts is equal to the number of partitions of n whose parts are all $\leq m$. For example, if $n = 5$ and $m = 3$, the partitions of the first type are 5, 4 + 1, 3 + 2, 3 + 1 + 1, 2 + 2 + 1, while those of the second type are 3 + 2, 3 + 1 + 1, 2 + 2 + 1, 2 + 1 + 1 + 1, 1 + 1 + 1 + 1 + 1.

b. Prove that if $n > m(m + 1)/2$, the number of partitions of n into m distinct parts is equal to the number of partitions of $n - m(m + 1)/2$ into at most m (not necessarily distinct) parts.

33a.* Prove that the number of partitions of any integer n into distinct parts is equal to the number of partitions of n into odd parts. For example, the partitions of 6 into distinct parts are 6, 5 + 1, 4 + 2, 3 + 2 + 1, while those into odd parts are 5 + 1, 3 + 3, 3 + 1 + 1 + 1, 1 + 1 + 1 + 1 + 1 + 1.

b. Prove that the number of partitions of n in which no integer occurs more than $k - 1$ times as a part is equal to the number of partitions of n into parts not divisible by k (Part a is the case $k = 2$). Thus if $k = 3$, $n = 6$, the partitions where no integer occurs more than twice among the parts are 6, 5 + 1, 4 + 2, 4 + 1 + 1, 3 + 3, 3 + 2 + 1, 2 + 2 + 1 + 1. The partitions in which no part is divisible by 3 are

---

[1] Some of Euler's results are contained in problems 53b, 145, 164.

$5 + 1, 4 + 2, 4 + 1 + 1, 2 + 2 + 2, 2 + 2 + 1 + 1, 2 + 1 + 1 + 1 +$
$1, 1 + 1 + 1 + 1 + 1 + 1.$

III. COMBINATORIAL PROBLEMS ON THE CHESSBOARD

The problems of this section involve various configurations of chess pieces on a chessboard. We will consider not only the usual chessboard of 8 rows and 8 columns, but also an $n \times n$ chessboard, having n rows and n columns. To understand these problems it is necessary to know the following:

A rook controls all squares of its row and column, up to and including the first square occupied by another piece.

A bishop controls all squares of the diagonals on which it lies up to and including the first square occupied by another piece.

The queen controls all squares of the row, column, and diagonals on which it lies, up to and including the first square occupied by another piece.[2]

The king controls all squares adjacent to the square on which it lies. (See fig. 2a; the square on which the king lies is marked with a circle and the squares controlled by the king are marked with crosses.)

A knight controls those squares which can be reached by moving one square horizontally or vertically and one square diagonally away from the square occupied by the knight. (See fig. 2b; the square occupied by the

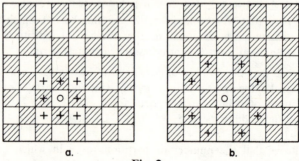

a.   b.

Fig. 2

[2] In accordance with what has been said, we count the square on which a rook, bishop, or queen lies as being controlled by it. In chess literature the square occupied by a piece is not considered to be controlled by that piece. To translate problems 34b, 35b, 36b, and 38 into the usual chess-player's language, the expression "every square of the board" in the hypotheses would have to be changed to "every unoccupied square of the board." (cf. hypothesis of problem 40.)

knight is marked by a circle and the squares the knight controls are marked by crosses.)

No other facts about the game of chess are necessary to understand and solve these problems.

34a. What is the greatest number of rooks which can be placed on an $n \times n$ chessboard in such a way that none of them controls the square on which another lies? In how many different ways can this be done?

b. What is the smallest number of rooks which can be arranged on an $n \times n$ chessboard in such a way that every square of the board is controlled by at least one of them? In how many different ways can this be done?

35a. What is the greatest number of bishops which can be arranged on an ordinary chessboard (8×8) in such a way that none of them controls the square on which another lies? Solve the same problem for an $n \times n$ chessboard.

b. What is the smallest number of bishops which can be arranged on an 8×8 chessboard in such a way that every square of the board is controlled by at least one bishop? Solve the same problem for an $n \times n$ chessboard.

36. Prove that for even n the following numbers are perfect squares:

a. the number of different arrangements of bishops on an $n \times n$ chessboard such that no bishop controls a square on which another lies and the maximum possible number of bishops is used.

b. the number of different arrangements of bishops on an $n \times n$ chessboard such that every square is controlled by at least one bishop and the minimum number of bishops is used.

37a.* Prove that in an arrangement of bishops which satisfies the hypotheses of problem 36a, the bishops all lie on the outermost rows or columns of the board.

b.** Determine the number of arrangements of bishops on an $n \times n$ board which satisfy the hypotheses of problem 36a.

38.** Determine the number of arrangements of bishops such that every square of the board is controlled by at least one bishop, and the smallest possible number of bishops is used:

    a. On an 8×8 chessboard.
    b. On a 10×10 chessboard.
    c. On a 9×9 chessboard.
    d. On an $n \times n$ chessboard.

39. What is the greatest number of kings which can be arranged in such a way that none of them lies on a square controlled by another

    a. On an 8×8 chessboard?
    b. On an $n \times n$ chessboard?

40. What is the smallest number of kings which can be arranged in such a way that every unoccupied square is controlled by at least one of them:
   a. On an 8 × 8 chessboard?
   b. On an $n \times n$ chessboard?

41. What is the greatest number of queens which can be arranged in such a way that no queen lies on a square controlled by another:
   a. On an 8 × 8 chessboard?
   b.*** On an $n \times n$ chessboard?

42a. What is the greatest number of knights which can be arranged on an 8 × 8 chessboard in such a way that none of them lies on a square controlled by another?
   b.** Determine the number of different arrangements of knights on an 8 × 8 chessboard such that no knight controls the square on which another lies, and the greatest possible number of knights is used.

   Some other combinatorial problems connected with arrangements of chess pieces can be found in L. Y. Okunev's booklet, *Combinatorial Problems on the Chessboard* (ONTI, Moscow and Leningrad, 1935).

IV. GEOMETRIC PROBLEMS INVOLVING COMBINATORIAL ANALYSIS

   Some of the problems in this group are concerned with convex sets. A set in the plane or in three-dimensional space is called *convex* if the line segment joining any two of its points is contained in the set. For example, the interior of a circle or of a cube is convex. The set S in fig. 3 is not convex, since the line segment joining A and B is not entirely contained in S.

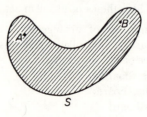

Fig. 3

43a. Each of the vertices of the base of a triangle is connected by straight lines to n points on the side opposite it. Into how many parts do these $2n$ lines divide the interior of the triangle?

b. Each of the three vertices of a triangle is joined by straight lines to n points on the opposite side of the triangle. Into how many parts do these $3n$ lines divide the interior of the triangle if no three of them pass through the same point?

44.* What is the greatest number of parts into which a plane can be divided by:
    a. n straight lines?
    b. n circles?

45.** What is the greatest number of parts into which three-dimensional space can be divided by:
    a. n planes?
    b. n spheres?

46.* In how many points do the diagonals of a convex n-gon meet if no three diagonals intersect inside the n-gon?

47.* Into how many parts do the diagonals of a convex n-gon divide the interior of the n-gon if no three diagonals intersect?

48. Two rectangles are considered different if they have either different dimensions or a different location. How many different rectangles consisting of an integral number of squares can be drawn
    a. On an 8×8 chessboard?
    b. On an $n \times n$ chessboard?

49. How many of the rectangles in problem 48 are squares
    a. On an 8×8 chessboard?
    b. On an $n \times n$ chessboard?

50.* Let K be a convex n-gon no three of whose diagonals intersect. How many different triangles are there whose sides lie on either the sides or the diagonals of K?

51.** *Cayley's problem.*[3] How many convex k-gons can be drawn, all of whose vertices are vertices of a given convex n-gon and all of whose sides are diagonals of the n-gon?

52. There are many ways in which a convex n-gon can be decomposed into triangles by diagonals which do not intersect inside the n-gon (see fig. 4, where two different ways of decomposing an octagon into triangles are illustrated).
    a. Prove that the number of triangles obtained in such a decomposition does not depend on the way the n-gon is divided, and find this number.

---

[3] Arthur Cayley (1821–1895), an English mathematician.

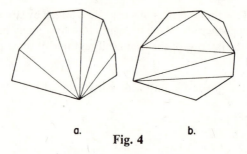

a.

b.

Fig. 4

b. Prove that the number of diagonals involved in such a decomposition does not depend on the way the n-gon is divided, and find this number.

53a. In how many different ways can a convex octagon be decomposed into triangles by diagonals which do not intersect within the octagon?

b.*** *Euler's problem*. In how many ways can a convex n-gon be decomposed into triangles by diagonals which do not intersect inside the n-gon?

54.*** $2n$ points are marked on the circumference of a circle. In how many different ways can these points be joined in pairs by n chords which do not intersect within the circle?

Problem 54 will reoccur later in another connection (see problem 84a). At that point some related problems (84b and 84c) will be given; for more general results, see the remark at the end of the solution of problem 84c.

55a. A circle is divided into p equal sectors, where p is a prime number. In how many different ways can these p sectors be colored with n given colors if two colorings are considered different only when neither can be obtained from the other by rotating the circle? (Note: It is not necessary that different sectors be of different colors or even that adjacent sectors be of different colors.)

b. Use the result of part a to prove the following *theorem of Fermat*[4]: If p is a prime number, then $n^p - n$ is divisible by p for any n.

56a. The circumference of a circle is divided into p equal parts by the points $A_1, A_2, \ldots, A_p$, where p is an odd prime number. How many different self-intersecting p-gons are there with these points as vertices if two p-gons are considered different only when neither of them can be

---

[4] Pierre Fermat (1601–1665), a French mathematician, was one of the creators of analytic geometry; he made many important contributions to number theory.

For other proofs of Fermat's theorem see, for example, L. E. Dickson, *Introduction to the Theory of Numbers* (U. of Chicago Press, 1929), p. 6 or G. H. Hardy and E. M. Wright, *An Introduction to the Theory of Numbers* (Oxford University Press, 1960), pp. 63–66.

obtained from the other by rotating the circle? (A self-intersecting polygon is a polygon some of whose sides intersect at other points besides the vertices; see, for example, the self-intersecting pentagons illustrated in fig. 5.)

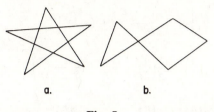

a.   b.

Fig. 5

b. Use the result of part a to prove the following *theorem of Wilson*[5]: If p is a prime number, then $(p - 1)! + 1$ is divisible by p.

V. PROBLEMS ON THE BINOMIAL COEFFICIENTS

The following problems will illustrate certain properties of the numbers

$$\binom{n}{k} = \frac{n!}{k!(n - k)!} = \frac{n(n - 1) \cdots (n - k + 1)}{1 \cdot 2 \cdots k} \qquad (0 \leq k \leq n) \quad (1)$$

In algebra courses it is proved that these numbers are the coefficients in the expansion

$$(1 + x)^n = \binom{n}{0} + \binom{n}{1}x + \binom{n}{2}x^2 + \cdots + \binom{n}{n}x^n$$

(the binomial theorem). In this connection the numbers $\binom{n}{k}$ are called the *binomial coefficients*. Using the binomial theorem one can obtain various relations involving the coefficients $\binom{n}{k}$; a direct proof of these relations from the formula (1) usually turns out to be appreciably more complicated than a proof using the binomial theorem.

[5] John Wilson (1741–1793), an English mathematician.
  For other proofs of Wilson's theorem, see Dickson, *op. cit.*, p. 15 or Hardy and Wright, *op. cit.*, pp. 68, 87.

57. Use the binomial theorem to evaluate the following sums:

a. $\binom{n}{0} + \binom{n}{1} + \binom{n}{2} + \cdots + \binom{n}{n}$

b. $\binom{n}{0} - \binom{n}{1} + \binom{n}{2} - \cdots + (-1)^n \binom{n}{n}$

c. $\binom{n}{0} + \frac{1}{2}\binom{n}{1} + \frac{1}{3}\binom{n}{2} + \cdots + \frac{1}{n+1}\binom{n}{n}$

d. $\binom{n}{1} + 2\binom{n}{2} + 3\binom{n}{3} + \cdots + n\binom{n}{n}$ $(n \geqq 1)$

e. $\binom{n}{1} - 2\binom{n}{2} + 3\binom{n}{3} - \cdots + (-1)^{n-1}n\binom{n}{n}$ $(n \geqq 1)$

f. $\binom{n}{0} - \binom{n}{1} + \binom{n}{2} - \cdots + (-1)^m\binom{n}{m}$ $(m \leqq n)$

g. $\binom{n}{k} + \binom{n+1}{k} + \binom{n+2}{k} + \cdots + \binom{n+m}{k}$ $(k \leqq n)$

h. $\binom{2n}{0} - \binom{2n-1}{1} + \binom{2n-2}{2} - \cdots + (-1)^n\binom{n}{n}$

i. $\binom{2n}{n} + 2\binom{2n-1}{n} + 4\binom{2n-2}{n} + \cdots + 2^n\binom{n}{n}$

j. $\binom{n}{0}^2 + \binom{n}{1}^2 + \binom{n}{2}^2 + \cdots + \binom{n}{n}^2$

k. $\binom{n}{0}^2 - \binom{n}{1}^2 + \binom{n}{2}^2 - \cdots + (-1)^n\binom{n}{n}^2$

l. $\binom{n}{0}\binom{m}{k} + \binom{n}{1}\binom{m}{k-1} + \binom{n}{2}\binom{m}{k-2} + \cdots + \binom{n}{k}\binom{m}{0}.$
   $(k \leqq \min(m, n))$

Some of the sums in this problem are encountered in another connection below, occasionally in a more general form. Thus, the sum of part l will be calculated by another method in the solution of problems 60a, 61c, and 72b. The result of part e will be generalized in the solution of problem 81c. The sum of part i will be determined by other means in the solution of problem 73b.

58. Use the binomial theorem to evaluate the following sums (the dots at the end of these sums indicate that the series are continued up to the point where the lower number becomes greater than the upper number):

a. $\dbinom{n}{0} + \dbinom{n}{2} + \dbinom{n}{4} + \dbinom{n}{6} + \cdots$

b. $\dbinom{n}{1} + \dbinom{n}{3} + \dbinom{n}{5} + \dbinom{n}{7} + \cdots$ $(n \geq 1)$

c. $\dbinom{n}{0} + \dbinom{n}{4} + \dbinom{n}{8} + \dbinom{n}{12} + \cdots$

d. $\dbinom{n}{1} + \dbinom{n}{5} + \dbinom{n}{9} + \dbinom{n}{13} + \cdots$ $(n \geq 1)$

e. $\dbinom{n}{2} + \dbinom{n}{6} + \dbinom{n}{10} + \dbinom{n}{14} + \cdots$ $(n \geq 2)$

f. $\dbinom{n}{3} + \dbinom{n}{7} + \dbinom{n}{11} + \dbinom{n}{15} + \cdots$ $(n \geq 3)$

g.* $\dbinom{n}{0} + \dbinom{n}{3} + \dbinom{n}{6} + \dbinom{n}{9} + \cdots$

h.* $\dbinom{n}{1} + \dbinom{n}{4} + \dbinom{n}{7} + \dbinom{n}{10} + \cdots$ $(n \geq 1)$

i.* $\dbinom{n}{2} + \dbinom{n}{5} + \dbinom{n}{8} + \dbinom{n}{11} + \cdots$ $(n \geq 2)$

59. *The factorial binomial theorem.* Let a and h be any real numbers, and n a positive integer. Let us introduce the notation:

$$a(a - h)(a - 2h) \cdots (a - (n - 1)h) = a^{n \mid h};$$

Thus, in particular, $a^{n \mid 0} = a^n$, $a^{1 \mid h} = a$. When $n = 0$ we define $a^{0 \mid h} = 1$. Prove that with this notation the following formula holds:

$$(a + b)^{n \mid h} = a^{n \mid h} + \dbinom{n}{1} a^{(n-1) \mid h} b^{1 \mid h} + \dbinom{n}{2} a^{(n-2) \mid h} b^{2 \mid h} + \cdots + b^{n \mid h}$$

This formula is called the *factorial binomial theorem.* It contains the ordinary binomial theorem as a special case (when $h = 0$).

60. Use the factorial binomial theorem to evaluate the following sums:

a. $\binom{n}{0}\binom{m}{k} + \binom{n}{1}\binom{m}{k-1} + \binom{n}{2}\binom{m}{k-2} + \cdots + \binom{n}{k}\binom{m}{0}$

b. $\binom{m}{0}\binom{n}{k} - \binom{m+1}{1}\binom{n}{k-1} + \binom{m+2}{2}\binom{n}{k-2} - \cdots$
$$+(-1)^k\binom{m+k}{k}\binom{n}{0}$$

Here $k \leq \min(m,n)$ in part a, and $k \leq n$ in part b.

In proving relations connecting the binomial coefficients it is sometimes helpful to make use of the fact that $\binom{n}{k}$ is the number of combinations of *n* objects taken *k* at a time (that is, the number of *k*-element subsets of a given set of *n* elements). To make such proofs more intuitively clear it is convenient to make use of the following geometric diagram. Suppose that we live in a town whose streets run in two perpendicular directions (see fig. 6, where all the streets of the town are represented in the form of horizontal

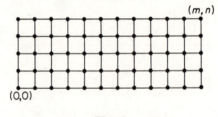

Fig. 6

and vertical lines). We can number the horizontal lines with the numbers 0, 1, 2, 3, $\cdots$ and do the same for the vertical lines. Then we can denote their intersections by pairs of coordinates (m,n), where *m* is the number of the "vertical" street which passes through the intersection and *n* is the number of the "horizontal" street (the intersections are denoted by dots in fig. 6). Suppose that we have to go from a house located at the intersection $(0,0)$ to a house located at the intersection (m,n). There will then be $\binom{m+n}{n}$ different shortest paths joining the two houses, for each of these shortest paths is $m + n$ blocks long—*m* blocks in the horizontal direction and *n* blocks in the vertical direction. A path is described unambiguously by specifying which of the $m + n$ blocks are the *n* vertical ones. One can choose which of the $m + n$ blocks are to be the *n* vertical ones in $\binom{m+n}{n}$ ways. By classifying the shortest paths in various ways, we can obtain with the aid of this diagram some interesting relations involving the binomial coefficients.

61a. Use the geometric scheme described above to prove that if $n \geq m$, then

$$\binom{n}{m} + \binom{n-1}{m-1} + \binom{n-2}{m-2} + \cdots + \binom{n-m}{0} = \binom{n+1}{m}.$$

b. Prove the following generalization of part a:

$$\binom{n}{m}\binom{k}{0} + \binom{n-1}{m-1}\binom{k+1}{1} + \binom{n-2}{m-2}\binom{k+2}{2} + \cdots$$
$$+ \binom{n-m}{0}\binom{k+m}{m} = \binom{n+k+1}{m},$$

where $n \geq m$.

c. Evaluate the sum

$$\binom{n}{0}\binom{m}{k} + \binom{n}{1}\binom{m}{k-1} + \binom{n}{2}\binom{m}{k-2} + \cdots + \binom{n}{k}\binom{m}{0}.$$
$$(k \leq \min(m, n))$$

62a. A network of roads is shown in fig. 7. 2^{1000} people leave the point A. Half go in the direction L and half in the direction R. Having reached the

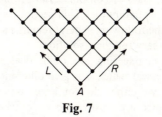

Fig. 7

first intersection, each group splits up, half going in the direction L and half in the direction R. The same thing happens at each subsequent intersection. How many people will reach each of the three leftmost intersections B_1, B_2, and B_3 of the thousandth row of intersections?

b. Solve this problem for all intersections of the thousandth row.

There are many more relations between the binomial coefficients. (See, for example, John Riordan, *An Introduction to Combinatorial Analysis*, Wiley, 1958, p. 14 ff.) Those readers who have solved all the problems of this section carefully will be able to set up many more such exercises.

Consider the following triangular array of numbers:

$$
\begin{array}{ccccccccc}
 & & & & 1 & & & & \\
 & & & 1 & 1 & 1 & & & \\
 & & 1 & 2 & 3 & 2 & 1 & & \\
 & 1 & 3 & 6 & 7 & 6 & 3 & 1 & \\
1 & 4 & 10 & 16 & 19 & 16 & 10 & 4 & 1
\end{array}
$$

In the uppermost (0-th) row of this triangle there is a single 1, and the numbers in the subsequent rows are determined by the following rule: each number is the sum of the three entries closest to it in the preceding row (that is, the sum of the number directly above it and the numbers immediately to the right and left of that number). In the n-th row of this array there are $2n + 1$ numbers; we will denote these numbers by B_n^0, B_n^1, B_n^2, ..., B_n^{2n}.

63.* Prove that

    a. $B_n^0 + B_n^1 + B_n^2 + \cdots + B_n^{2n} = 3^n$;

    b. $B_n^0 - B_n^1 + B_n^2 - \cdots + B_n^{2n} = 1$;

    c. $(B_n^0)^2 + (B_n^1)^2 + (B_n^2)^2 + \cdots + (B_n^{2n})^2 = B_{2n}^{2n}$.

After solving problem 63 the reader will be able to set up other relations involving the numbers B_n^k.

VI. PROBLEMS ON COMPUTING PROBABILITIES

A very important class of combinatorial problems is concerned with the computation of probabilities. This section is devoted to some of these problems, and the following general remarks are intended to provide the background necessary for their solution.

In science and engineering we often deal with experiments (or observations or processes) which can give different results depending on circumstances which we either do not know or are unable to control. For example, when dice are thrown we cannot know beforehand what numbers will come up, since this depends on circumstances not entirely within our control (the details of the motion of the hand in throwing the dice, particulars about the surface on which the dice fall, etc.). Similar remarks apply to the tossing of a coin or the spinning of a roulette wheel.

Let A be a given outcome of such an experiment, and suppose that the experiment is performed n times. Then A will be the outcome a certain number n_A of times, where $0 \leq n_A \leq n$. If the ratio n_A/n approaches a limit p as n becomes indefinitely large, we say that p is the *a posteriori probability* of the occurrence of A. Thus p represents the limiting value of the frequency with which A occurs in a series of trials of the experiment. We often write $p = Pr\{A\}$; from the definition we see that $0 \leq Pr\{A\} \leq 1$.

For example if the experiment consists of throwing a die, and if A is the outcome that a 5 turns up, then $Pr\{A\} = \frac{1}{6}$, since in a long series of throws the frequency with which a 5 turns up approaches $\frac{1}{6}$. If the experiment is to toss a coin, and if A denotes "heads," then $P\{A\} = \frac{1}{2}$. The probability of an absolute certainty is 1; thus one can say that the day after Saturday will be Sunday "with a probability of 1." The probability of an impossible event is 0; thus the probability that a 100 will turn up when a die is thrown is 0.

Very often we want to determine the probability that the result of an experiment is in a given set of outcomes. For example we might wish to know the probability that an even number will turn up when a die is thrown. In this case the given set of outcomes consists of 2, 4, and 6. Such a set E is called an *event*; if the outcome of the experiment is in E, we say that the event E has *occurred*. The a posteriori probability of an event E is defined in the same way as for a single outcome; we perform the experiment n times, let n_E denote the number of times that E occurred, and define $Pr\{E\}$ to be the limit of n_E/n as n becomes indefinitely large. In the above example where $E = \{2,4,6\}$ we have $Pr\{E\} = \frac{1}{2}$.

Let E_1 and E_2 be two events and denote by $E_1 \cup E_2$ (read "E_1 or E_2") the event obtained by combining the outcomes of E_1 and E_2 into a single set. It follows from the definition of probability that if E_1 and E_2 are *disjoint*, i.e. have no outcomes in common, then

$$Pr\{E_1 \cup E_2\} = Pr\{E_1\} + Pr\{E_2\}. \tag{1}$$

For example in the experiment of throwing a die, let $E_1 = \{5\}$ and $E_2 = \{2,4,6\}$. Then E_1 and E_2 are disjoint, $E_1 \cup E_2 = \{2,4,5,6\}$ and $Pr\{E_1 \cup E_2\} = \frac{1}{6} + \frac{1}{2} = \frac{2}{3}$.

In the case of tossing a coin or throwing a die, we feel intuitively that it is not necessary to carry out the experiment in order to determine the probabilities of various events. This is because of the symmetry which makes a head as likely as a tail and one face of a die as likely as any other. This idea leads naturally to the notion of *a priori probability*. If an experiment has a finite number N of possible outcomes, and if there is some symmetry or other reason present to make us consider these outcomes as "equally likely," then we assign the a priori probability $1/N$ to each of them. Thus $N = 6$ in the case of the die, and each face has an a priori probability of $\frac{1}{6}$. The a priori probability of an event E is then defined as N_E/N, where N_E is the number of outcomes in the set E. When calculating this quantity it is often convenient to refer to the outcomes in E as *favorable* and those not in E as *unfavorable*. Then we can say that the a priori probability of E is the number of favorable outcomes divided by the total number of outcomes. But it should be emphasized that this is true only when the N outcomes of the experiment are equally likely.

In cases where the a priori probability of E exists, it is equal to the a posteriori probability, for otherwise the assumption that the N outcomes were equally likely was erroneous.[6] All the problems in this section are to be worked using a priori probabilities.

[6] For a more philosophical discussion of this point, see H. Reichenbach, *The Theory of Probability*, University of California Press, Berkeley and Los Angeles, 1949.

Now let E and F be any two events, and denote by $E \cap F$ (read "E and F") the event consisting of all outcomes which are in both E and F. For example, in the experiment of throwing a die, if $E = \{1,2,4,6\}$ and $F = \{2,3,5,6\}$, then $E \cap F = \{2,6\}$. In the case where $Pr\{F\} > 0$ we denote the quantity $Pr\{E \cap F\}/Pr\{F\}$ by $Pr\{E \mid F\}$ and call it the *conditional probability of E given F*. To understand the meaning of this quantity consider the case of an experiment with N equally likely outcomes, N_F of them in F and $N_{E \cap F}$ of them in $E \cap F$. Then

$$Pr\{E \mid F\} = \frac{N_{E \cap F}/N}{N_F/N} = \frac{N_{E \cap F}}{N_F}.$$

Thus $Pr\{E \mid F\}$ is the fraction of the outcomes of F which are also in E. If we select at random an outcome of F, $Pr\{E \mid F\}$ is the probability that it will be in E.

For example if a die is thrown and if an even number is known to have come up, what is the probability that it was a multiple of 3? Here $E = \{3,6\}$, $F = \{2,4,6\}$, $E \cap F = \{6\}$, and so $Pr\{E \mid F\} = \frac{1}{6}/\frac{1}{2} = \frac{1}{3}$.

From the definition of $Pr\{E \mid F\}$ we see that

$$Pr\{E \cap F\} = Pr\{F\}\, Pr\{E \mid F\}.$$

In words: the probability of $E \cap F$ is the probability of F times the conditional probability of E given F.

If two events E and F are such that

$$Pr\{E \cap F\} = Pr\{E\}\, Pr\{F\},$$

we say they are *independent* of each other. To see the meaning of this concept, suppose $Pr\{F\} > 0$; then $Pr\{E \mid F\} = Pr\{E \cap F\}/Pr\{F\} = Pr\{E\}Pr\{F\}/Pr\{F\} = Pr\{E\}$. Thus the conditional probability of E given F is the same as the (absolute) probability of E. In other words, the probability of the occurrence of E is not changed by a knowledge of whether of not F occurred.

For example, consider an experiment which consists of tossing a coin twice. Let E be the event that a head comes up on the first toss, and let F be the event that a head comes up on the second toss. Then $E \cap F$ is the event that heads come up on both tosses. We have $Pr\{E \cap F\} = \frac{1}{4} = Pr\{E\}Pr\{F\}$, and so E and F are independent. In most applications of this concept we know $Pr\{E\}$ and $Pr\{F\}$ and also that E and F are independent; we then apply the formula $Pr\{E \cap F\} = Pr\{E\}Pr\{F\}$ to evaluate $Pr\{E \cap F\}$.

More generally we say that the events $E_1, E_2, \ldots, E_n$ are independent if

$$Pr\{E_1 \cap E_2 \cap \cdots \cap E_n\} = Pr\{E_1\}Pr\{E_2\} \cdots Pr\{E_n\},$$

where $E_1 \cap E_2 \cap \cdots \cap E_n$ consists of the outcomes which are in all of the events $E_1, E_2, \ldots, E_n$.

Now consider an experiment with N equally likely possible outcomes, and let E and F be any two events. Denote by N_E, N_F, $N_{E \cup F}$, and $N_{E \cap F}$ the numbers of outcomes in the events E, F, $E \cup F$, and $E \cap F$ respectively Then

$$N_{E \cup F} = N_E + N_F - N_{E \cap F}$$

(see the solution to problem 12a). Dividing by N we obtain

$$Pr\{E \cup F\} = Pr\{E\} + Pr\{F\} - Pr\{E \cap F\}.$$

This formula is often useful in computing probabilities; by applying problem 12c a similar formula can be derived for $Pr\{E_1 \cup E_2 \cup \cdots \cup E_n\}$, where $E_1 \cup E_2 \cup \cdots \cup E_n$ denotes the event obtained by combining the outcomes of $E_1, E_2, \ldots, E_n$ in a single set.[7]

64. In a certain town there are 10,000 bicycles, each of which is assigned a license number from 1 to 10,000 (no two bicycles receive the same number). What is the probability that the number on the first bicycle one encounters will not have any 8's among its digits?

65a. Six cards bearing respectively the letters A, B, C, D, E, and F are shuffled thoroughly, and then the top four cards are turned face up (without changing the order in which they lay on top of the shuffled pack). What is the probability that they will spell out the word "DEAF"?

b. The same process is performed on a set of cards consisting of three D's, two O's and one X. What is the probability that the top four cards will spell out the word "DODO"?

66.* Ten slips of paper bearing the numbers 0, 1, 2, 3, 4, 5, 6, 7, 8, and 9 are put into a hat. Five slips are drawn at random and laid out in a row in the order in which they were drawn. What is the probability that the five-digit number thus formed will be divisible by 495?

67. Suppose that a boy remembers all but the last figure of his girl friend's telephone number and decides to choose the last figure at random in an attempt to reach her. If he has only two dimes in his pocket, what is the probability that he will dial the right number before he runs out of money?

68. For the purposes of this problem, suppose that the probability that a person's birthday falls in any given month is 1/12. What is the probability that:

a. in a given group of 12 people, no two of them celebrate their birthdays in the same month?

b. the birthdays of 6 given people all fall in only two different

---

[7] The above discussion of probability theory is of course only the briefest of introductions; for further information the reader is referred to the following books: W. Feller, *An Introduction to Probability Theory and its Applications*, Wiley, New York, 1950; E. Parzen, *Modern Probability Theory and its Applications*, Wiley, New York, 1960; H. Cramér, *The Elements of Probability Theory*, Wiley, New York, 1955.

months? (This means that they must *not* all be born in the *same* month.)

69. Nine passengers board a train consisting of three cars. Each passenger selects at random which car he will sit in. What is the probability that:

    a. there will be three people in the first car?

    b. there will be three people in each car?

    c. there will be two people in one car, three in another, and four in the remaining car?

70. A pack of ten cards, numbered from 1 to 10, is shuffled and dealt into two five-card hands.

    a. What is the probability that the 9 and 10 are in the same hand?

    b. What is the probability that the 8, 9 and 10 are all in the same hand?

    c. What is the probability that of the four highest cards, two are in one hand and two in the other?

71. Suppose A and B are two equally strong ping-pong players. Is it more probable that A will beat B in 3 games out of 4, or in 5 games out of 8?

72a. k balls are selected at random from a box containing n white balls and m black ones. What is the probability that exactly r of the balls drawn are white?

    b. Apply the result of part a to evaluate the sum

$$\binom{n}{0}\binom{m}{k} + \binom{n}{1}\binom{m}{k-1} + \binom{n}{2}\binom{m}{k-2} + \cdots + \binom{n}{k}\binom{m}{0}.$$

Remark. For other methods of determining this sum, see the solutions to problems 57l, 60a, 61c.

73a. *Banach's matchbox problem.*[8] A man buys two boxes of matches and puts them in his pocket. Every time he has to light a match, he selects at random one box or the other. After some time the man takes one of the boxes from his pocket, opens it, and finds that it is empty. (Note: the man must then have absentmindedly put the empty box back in his pocket after he had used the last match in it.) What is the probability that there are at that moment k matches left in the other box if each box originally contained n matches? Here $0 \leqq k \leqq n$.

    b. Use the result of part a to evaluate the sum

$$\binom{2n}{n} + 2\binom{2n-1}{n} + 2^2\binom{2n-2}{n} + \cdots + 2^n\binom{n}{n}$$

Remark. For another method of determining this sum, see the solution to problem 57i.

74.* Two hunters A and B set out to hunt ducks. Each of them hits as

---

[8] Stephen Banach (1892–1945), a Polish mathematician.

often as he misses when shooting at ducks. Hunter A shoots at 50 ducks during the hunt and hunter B shoots at 51. What is the probability that B bags more ducks than A?

75a. Two hunters see a fox and shoot at it simultaneously. Assume that each of the hunters averages one hit per three shots. What is the probability that at least one of the hunters will hit the fox?

b. Solve the same problem for the case of three hunters, assuming that the accuracy of each of the three hunters is the same as in part a.

c. Solve the same problem for the case of n hunters.

76. A hunter shoots from a distance of 100 yards at a fox running away from him; suppose that the probability that he hits it at this distance is 1/2 (that is, from a distance of 100 yards the hunter hits a running fox just as often as he misses). If he misses, the hunter reloads his rifle and shoots again, but in the time it takes to do this the fox runs 50 yards. If he misses a second time, he reloads the rifle and shoots a third (and last) time, the fox having meanwhile run another 50 yards. Under the hypothesis that the probability of a hit is inversely proportional to the square of the distance, determine the probability that the hunter succeeds in hitting the fox.

77. *The problem of the four liars.* It is known that each of four people, A, B, C, and D, tells the truth in only one case out of three. Suppose that A makes a statement, and then D says that C says that B says that A was telling the truth. What is the probability that A was actually telling the truth?

Remark. This problem can also be formulated in the following way. A slip of paper is given to A, who marks it with either a plus or a minus sign; the probability of his writing a plus is known to be 1/3. He then passes the slip to B, who may either leave it alone or change the sign before passing it on to C. Next C passes the slip to D after perhaps changing the sign; finally D passes it to an honest judge after perhaps changing the sign. The judge sees a plus sign on the slip. It is known that B, C, and D each change the sign with probability 2/3. What is the probability that A originally wrote a plus?

78a. In certain rural areas of Russia fortunes were once told in the following way. A girl would hold six long blades of grass in her hand with the ends protruding above and below; another girl would tie together the six upper ends in pairs and then tie together the six lower ends in pairs. If it turned out that the girl had thus tied the six blades of grass into a ring, this was supposed to indicate that she would get married within a year. What is the probability that a ring will be formed when the blades of grass are tied at random in this fashion?

b. Solve the same problem for the case of $2n$ blades of grass.

79a.* A jar contains $2n$ thoroughly mixed balls, n white and n black.

What is the probability that each of n people drawing two balls blind-folded from the jar will draw balls of different colors? (The balls drawn are not replaced in the jar.)

b. Under the same conditions, what is the probability that each of the n people draws two balls of the same color?

80a.*** An absent-minded professor wrote n letters and sealed them in envelopes without writing the addresses on the envelopes. Having forgotten which letter he had put into which envelope, he wrote the n addresses on the envelopes at random. What is the probability that at least one of the letters was addressed correctly?

b. What limit does the probability of part a approach as $n \to \infty$?

81a.** A train consists of n carriages. Each of p passengers selects at random the carriage in which he will ride. What is the probability that there will be at least one passenger in each carriage?

b. Under the hypotheses of part a, what is the probability that exactly r of the carriages will be occupied?

c. Use the result of part a to evaluate the sum

$$\binom{n}{1}1^p - \binom{n}{2}2^p + \binom{n}{3}3^p - \cdots + (-1)^{n-1}\binom{n}{n}n^p,$$

where $1 \leq p \leq n$.

Remark. Problem 81b is equivalent to the following problem, which is of interest to the physicist: a stream of p particles is caught by a system of n receptors of some particle-counting apparatus. Each particle is equally likely to hit any given receptor. What is the probability that particles will hit exactly r of the receptors?

The computation of the sum of problem 81c for the special case of $p = 1$ was treated above in problem 57e.

82.** The twenty letters a, b, c, d, e, f, g, h, i, j, A, B, C, D, E, F, G, H, I, J are written down on separate slips of paper; the ten slips bearing the capital letters are shuffled and then arranged in random order in a circle; then the ten slips with small letters are shuffled and placed at random in the spaces between the first ten slips. What is the probability that no small letter will be adjacent to the corresponding capital letter?

83a.*** $n + m$ people are waiting in line at a box office; n of them have five-dollar bills and the other m have nothing smaller than ten-dollar bills. The tickets cost \$5 each. When the box-office opens there is no money in the till. If each customer buys just one ticket, what is the probability that none of them will have to wait for change?

b. Solve the same problem under the assumption that initially there were p five-dollar bills in the till.

c. For the purposes of this problem, assume that there exist three-dollar bills. $n + m$ people are standing in line at a box-office; n of them have single dollars and the other m of them only have three-dollar bills. The tickets cost \$1 each and each person wants one ticket. When the box office opens there is no money in the till. What is the probability that none of the customers will have to wait for change?

Remark. Problems 83a–c, despite their artificial formulation, are of great interest in practical applications; certain problems in physics and in the theory of statistical control of production lead to such situations.

84a.* Use the results of problem 83a to derive a new solution to problem 54.

b. $3n$ points are marked on the circumference of a circle. In how many ways can they be divided into n sets of three in such a way that no two of the inscribed triangles determined by these sets of three points intersect each other?

c. In how many ways can a convex $2n$-gon be decomposed into quadrilaterals by drawing diagonals which do not meet inside the $2n$-gon?[9]

85a. In a card game there are $m + n$ players and a banker who does not play but only collects and distributes money. At the beginning of the game there is no money in the bank. A pack consisting of m cards marked "win" and n cards marked "lose" is dealt, each player receiving one card. The first player then turns over his card; if it is a winner he collects a dollars from the bank, but if it is a loser he pays b dollars to the bank. Then the second player turns over his card, and so on. Assume that the total amount won is equal to the total amount lost; i.e., $ma = nb$. Suppose also that m and n are relatively prime. What is the probability that throughout the game the banker always has enough money on hand to pay the winners?

b. What is the probability that at exactly k stages of the game there is a negative amount of money in the bank? Here k is an integer in the range $0 \leqq k \leqq m + n - 1$.

VII. EXPERIMENTS WITH INFINITELY MANY POSSIBLE OUTCOMES

In the preceding section we dealt with experiments having a finite number of equally likely possible outcomes. In that case the probability

---

[9] It is not hard to see that a polygon with an odd number of sides can never be decomposed into quadrilaterals as required in the problem.

of an event is the number of favorable outcomes divided by the total number of possible outcomes. There are however, cases in which neither the number of possible outcomes to the experiment nor the number of them in which the event takes place is finite, but nevertheless the notion of probability can be given a definite meaning which allows one to compute it with the aid of combinatorial considerations. Thus, for example, there are infinitely many positive integers. However, the question of determining the probability that a positive integer selected at random is divisible by 5 still makes sense; most people would say that this probability is 1/5, even though we have as yet given no definition applicable to this case.

To formulate such a definition, consider the following more general problem. Let there be given an infinite sequence of numbers

$$a_1, a_2, a_3, \ldots$$

Suppose that the first N of these numbers are written on N slips of paper, the slips thoroughly mixed, and then one of them drawn at random. This experiment has N equally probable outcomes; if we denote by $q(N)$ the number of members of the sequence $a_1, a_2, a_3, \ldots, a_N$ which possess some given property, then the probability that the slip drawn bears a number possessing this property is $q(N)/N$.

Suppose that as $N \to \infty$ the ratio $q(N)/N$ approaches a limit; in this case this limit is called the probability that a number selected at random from the entire sequence has the desired property.

Note that this probability depends on the way in which the numbers are arranged in a sequence. Changing the order of the numbers can change the value of the probability. Example: consider the positive integers arranged in increasing order: 1, 2, 3, ... Of the first N of these numbers, $[N/2]$ are even; as $N \to \infty$ the ratio $[N/2]/N$ approaches $\frac{1}{2}$, which means that the probability that any number selected at random is even equals 1/2. Now let the positive integers be arranged in the order 1, 3, 2, 5, 7, 4, 9, 11, 6, ..., that is, the first two odd numbers, then the first even number, then the next two odd numbers, then the next even number, etc. Among the first N of these numbers, there are only $[N/3]$ even numbers, and as $N \to \infty$ the ratio $[N/3]/N$ approaches the limit 1/3, thus making the probability that any number selected at random is even equal to 1/3. In the problems below, it is assumed that the positive integers are arranged in the order 1, 2, 3, ... ; but always keep in mind that if they were arranged in a different order, a different result might be obtained.

If the sequence consists of the positive integers arranged in increasing order and the property in question is that of divisibility by 5, the above definition leads to a probability of 1/5. To see this, note first that $q(N) = [N/5]$. Now any number N can be written in the form $N = 5q + r$, where $q = [N/5]$ and r is the remainder upon division of N by 5 (and thus equal to 0, 1, 2, 3, or 4). It follows from this that

$$\frac{[N/5]}{N} = \frac{q}{N} = \frac{N-r}{5N} = \frac{1}{5} - \frac{r}{5N}$$

and, since $0 \leq r \leq 4$,

$$\lim_{N \to \infty} \frac{[N/5]}{N} = \lim_{N \to \infty} \left(\frac{1}{5} - \frac{r}{5N} \right) = \frac{1}{5}.$$

Thus when N is a very large (but finite) positive integer, the probability that a number selected at random from the first N positive integers will be divisible by 5 is very close to 1/5. The probability that the serial number on a one-dollar bill chosen at random will be divisible by 5 is nearly equal to 1/5 (since the total number of one-dollar bills in circulation is very great); it is not necessary to know the exact number of dollars in circulation. We expect the reader to make this approximation in problems 86–94 below. The hypotheses of these problems may seem artificial, but there are problems of the same type whose solutions are of practical importance.

86. What is the probability that a positive integer selected at random is relatively prime to 6? That at least one of two integers selected at random is relatively prime to 6?

87a. What is the probability that the square of an integer selected at random will end with the digit 1? That the cube of an integer selected at random will end with the digits 11?

   b. What is the probability that the final digit of the tenth power of a number selected at random is a 6? That the final digit of the twentieth power of a number selected at random is a 6?

88. What is the probability that when n is selected at random from the positive integers greater than 7: **a.** $\binom{n}{7}$ is divisible by 7? **b.** $\binom{n}{7}$ is divisible by 12?

89. What is the probability that the final digit of 2^n, where n is a positive integer selected at random, is a 2? That the last two digits are 12?

90.* What is the probability that the first digit of 2^n is a 1?

91a.*** Prove that 2^n can begin with any sequence of digits.

   b. Let M be any k-digit number. What is the probability that the first k digits of the number 2^n represent the number M?

   Remark. Problem 90 is a special case of problem 91b.

92.** Let N be a positive integer, and let s_N be the probability that two integers a, b, chosen at random from the range $1 \leq a, b \leq N$, are relatively prime. Prove that

$$\lim_{N \to \infty} s_N = s$$

exists.

93. Show that the infinite series

$$1 + \frac{1}{2^2} + \frac{1}{3^2} + \frac{1}{4^2} + \cdots$$

converges to the value $1/s$, where s is the number defined in problem 92. Using this, compute s numerically to within 0.1.

94.** Prove that the probability that an integer is prime is 0. In other words prove that if $\pi(N)$ denotes the number of primes $\leq N$, then

$$\lim_{N \to \infty} \frac{\pi(N)}{N} = 0.$$

VIII. EXPERIMENTS WITH A CONTINUUM OF POSSIBLE OUTCOMES

To conclude this volume, we consider some more problems on computing probabilities, this time involving experiments whose possible outcomes can be represented by the points of a line segment, or of some plane figure, or solid body. In such cases one cannot speak of the number of outcomes in which a given event occurs; nevertheless, one can often define the probability of the event in a natural way and calculate it by geometrical considerations. The easiest way to explain how such computations are performed is through concrete examples.

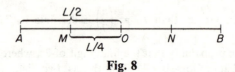

Fig. 8

Example 1. A rod of length L is broken at a point chosen at random. What is the probability that the smaller of the two pieces will be of length greater than $L/4$?

The possible outcomes of the experiment correspond to the different points at which the rod can be broken, that is, the set of all possible outcomes to the experiment can be represented as the totality of all points on a segment AB of length L (fig. 8).

Now exactly what is meant by saying that the rod is broken at a point "chosen at random"? If we stipulate that all points of AB have the same probability p of being chosen, then we must have $p = 0$, since there are infinitely many such points. This does not give sufficient information to calculate the probability required by the problem. What we must do is to associate to each *interval CD* on the line *AB* a number $p(CD)$, the probability that the break point lies between *C* and *D*. Since $p(CD)$ is to be interpreted as a probability, it must satisfy the inequalities

$$0 \leqq p(CD) \leqq 1.$$

Since the break point is certain to be between *A* and *B*, we must have

$$p(AB) = 1.$$

If *C*, *D*, and *E* are three points on the line as in fig. 9 we will require that

$$p(CE) = p(CD) + p(DE).$$

This requirement is a natural extension of property (1) on page 21. We can now define the phrase "at random" precisely; we will use this term to mean that the probability $p(CD)$ depends only on the *length* of *CD* and not on its *location* on the rod.

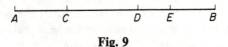

Fig. 9

In this case we can write $p(x)$ instead of $p(CD)$, where x denotes the length of the segment *CD*. The function $p(x)$ is defined for all values of x in the range $0 \leqq x \leqq L$. The above properties can now be written in the form

(1) $0 \leqq p(x) \leqq 1$

(2) $p(L) = 1$

(3) $p(x + y) = p(x) + p(y)$ if $x + y \leqq L$.

We will now show that there is only one function $p(x)$ having these three properties, namely $p(x) = x/L$. We first note that property (3) can be generalized to

(4) $p(x_1 + x_2 + \cdots + x_n) = p(x_1) + p(x_2) + \cdots + p(x_n)$

$$\text{if } x_1 + x_2 + \cdots + x_n \leqq L.$$

For example, if $n = 3$, we can apply property (3) twice to obtain

$$p(x_1 + x_2 + x_3) = p((x_1 + x_2) + x_3)$$
$$= p(x_1 + x_2) + p(x_3) = p(x_1) + p(x_2) + p(x_3).$$

The general case follows readily by mathematical induction.

In (4) put $x_1 = x_2 = \cdots = x_n = L/n$. Then $x_1 + x_2 + \cdots + x_n = L$, and so by (2), $p(x_1 + \cdots + x_n) = 1$. Hence (4) becomes $1 = np(L/n)$, or

$p(L/n) = 1/n$. Next suppose that m, n are positive integers with $m \leqq n$. By (4) we have

$$p\left(\frac{m}{n} L\right) = p\left(\overbrace{\frac{L}{n} + \frac{L}{n} + \cdots + \frac{L}{n}}^{m \text{ terms}}\right)$$

$$= p\left(\frac{L}{n}\right) + p\left(\frac{L}{n}\right) + \cdots + p\left(\frac{L}{n}\right)$$

$$= mp\left(\frac{L}{n}\right) = \frac{m}{n}.$$

This means that $p(x) = x/L$ whenever x/L is a rational number. To deal with the case where x/L is irrational, note first that if $0 \leqq x \leqq y \leqq L$, then

$$p(y) = p(x) + p(y - x) \geqq p(x),$$

since $p(y - x) \geqq 0$ by property (1). Thus the function $p(x)$ is *monotone non-decreasing*. Now if x/L is irrational, and n is any positive integer, we can choose rational numbers a/L and b/L such that $a \leqq x \leqq b$ and such that $b/L - a/L < 1/n$. (This is because the rational numbers are everywhere dense, so that any irrational number can be approximated arbitrarily closely by rational numbers.) We then obtain

$$\frac{a}{L} = p(a) \leqq p(x) \leqq p(b) = \frac{b}{L},$$

from which it follows that $|p(x) - x/L| < 1/n$. Since this holds for all n, we see that $p(x) = x/L$.

Conversely, it is easily verified that the function $p(x) = x/L$ actually satisfies (1)–(3).

Thus when the rod AB is broken at random, the probability that the break point lies in an interval CD is equal to the length of CD divided by the length of AB, i.e. the fraction of the total length contained in the interval CD.

To return to the original problem, we note that the length of the smaller of the pieces into which the rod is broken will be greater than $L/4$ if and only if the break point lies within the segment MN (see fig. 8) whose length is $L/2$, and whose endpoints lie at a distance of $L/4$ from the ends of the rod. Hence the required probability is $(L/2)/L = \frac{1}{2}$.

Example 2. A coin of diameter d is tossed on a tiled floor. The tiles of the floor are squares of width $a > d$. What is the probability that the coin will lie entirely within one of the squares (that is, that it will not cross any of the sides of a square)?

The possible outcomes of the experiment considered in this problem correspond to the various points at which the center of the coin can land.

Since all the squares on the floor are congruent to each other, it will suffice to consider a single square. Let us restrict ourselves for the moment to the cases in which the center of the coin lands within the square under consideration. The set of all possible outcomes to the experiment can be represented by the totality of all points within a square $ABCD$ of width a.

The term "at random" in the formulation of this problem is to be understood as meaning that the probability that the center of the coin will land in a given rectangle depends on the area of this rectangle and not on its location within the square $ABCD$. By a proof analogous to that given on pages 31–32, it can be shown that this implies that the probability of the center of the coin landing in any given rectangle is the

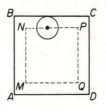

Fig. 10

ratio of the area of that rectangle to the area of the entire square $ABCD$. But it is easy to see that the coin will lie entirely within the square $ABCD$ if and only if its center lands within the square $MNPQ$ of width $a - d$ whose center coincides with that of the square $ABCD$ and whose sides are parallel to those of $ABCD$ (fig. 10). It follows from this that the required probability is

$$\frac{\text{area } MNPQ}{\text{area } ABCD} = \frac{(a - d)^2}{a^2} = \left(1 - \frac{d}{a}\right)^2.$$

Suppose that the floor consists of n squares. We have divided the outcomes into n classes: one for each square, consisting of the outcomes in which the coin's center lands within that square. If it is equally probable that the center of the coin will land in any of the n squares and if the above argument is valid for each of the squares, then the probability of the coin lying entirely within a given square $(1/n)(1 - d/a)^2$ (since $pr\{E \cap F\} = pr\{E\}pr\{F\}$ when E and F are independent events). By adding up these probabilities for all n squares, we obtain

$$n \cdot \frac{1}{n}\left(1 - \frac{d^2}{a}\right) = \left(1 - \frac{d}{a}\right)^2$$

as the probability that the coin will lie entirely within one of the squares. Note that the argument might fail to be valid for the edge squares of the floor (for example, if the tiles run wall-to-wall, the center of the coin could not lie closer than $d/2$ to the outer edge of an edge square). However, if the number of interior squares on the floor greatly exceeds the number of edge squares, the error involved in using the above result is negligible. The error disappears altogether for a floor of infinite length and width.

Note that in the examples considered, the solution to the problem depended heavily on the interpretation of the term "at random." In all cases where any doubt can arise as to the meaning of this term, it is necessary to make precise in the statement of the problem how the term is to be understood; without this the problem will not have an unambiguous solution (see the remark below). In the following problems, however, such doubt can scarcely arise: in all of them it is possible to define "at random" in almost the same way as in the above examples, and this is the way the term is to be understood throughout.

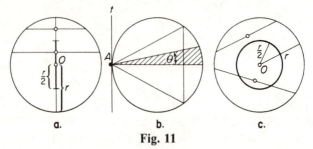

Fig. 11

Remark. A classical example of a problem in which there are many different possible meanings for the term "at random," each of them making sense intuitively, but each giving rise to different values for the probability is the following:

"What is the probability that a chord drawn at random on a circle will be longer than the side of the inscribed equilateral triangle?"

Various different answers can be obtained by interpreting the term "at random" in various ways. For example, because of the symmetry of the figure, we can restrict our attention to chords parallel to a given direction and interpret the term "at random" as follows (fig. 11a). Draw the diameter which is perpendicular to the given direction; each chord is uniquely determined by specifying the point at which it meets this diameter. Interpret the term "at random" as meaning that the probability of this intersection point lying within a given segment of the diameter shall depend only on the length of that segment The desired probability will then be $1/2$, since the chords longer than a side of the inscribed equilateral triangle are those which lie at a distance of less than $r/2$ from the center.

On the other hand, again because of symmetry, we can restrict our attention to those chords which pass through a given point A of the circumference (fig. 11b). Let t be the tangent to the circle at A. A chord can be thought of as a ray emanating from A and lying on the same side of t as the circle. Let us interpret the term "at random" as meaning that the probability that this ray lies within a given angle θ depends only on the magnitude of θ and not on its position (as long as θ is on the same side of t as the circle.) Since the angles of an equilateral triangle are 60°, it then follows that the desired probability is $60°/180° = 1/3$.

Finally, one can also interpret the term "at random" to mean that the probability that the midpoint of a chord lies within a given circle is proportional to the area of that circle. (Note that the midpoint of a chord uniquely determines the chord, unless it is a diameter.) Since any point inside the circle can be the midpoint of a chord, and the points which are midpoints of chords longer than a side of the inscribed equilateral triangle form the interior of a circle of radius $r/2$ (see fig. 11c), the desired probability must, in accordance with this interpretation of the term "at random," have the value $\pi(r/2)^2/\pi r^2 = \frac{1}{4}$.

This example emphasizes the fact that a proper use of the term *probability* requires the specification of a definite experiment. The term "at random" must be understood as an indication of just how the experiment in question is to be performed. In the above examples about breaking the rod and tossing the coin onto the tile floor, and in problems 95–100 below, the sense of this word is clear from the statement of the problem. In the case of a "random" selection of a chord on a circle it is not clear without any further explanation just how this selection is to be carried out, and the term "at random" gives no explanation by itself. If we draw a circle on a large sheet of paper, drop a needle onto the paper and (in those cases where the point of the needle lies within the circle) draw a chord whose midpoint is the point of the needle; then the third of our interpretations is valid and the desired probability is $\frac{1}{4}$. If we select a point A on the circumference of the circle, fasten a rod at A, spin it about A, and draw a chord in the direction in which the rod comes to rest, then the second interpretation is valid (neglecting friction) and the desired probability is $\frac{1}{3}$. Finally, it is not hard to show that for most of the more natural ways of selecting a chord "at random" (dropping a circular disk onto a plane surface on which a line has been drawn, dropping a long rod onto a plane surface on which a circle has been drawn, observing the trajectory of stars passing behind the moon or of particles moving linearly in the [circular] field of vision of a microscope or telescope), the first of the three interpretations is valid and the required probability is $\frac{1}{2}$; in this sense, the first of the three solutions presented is the "most correct."

95. Two people agree to meet at a given place between noon and 1 PM. By agreement, the first to arrive will wait 15 minutes for the second, after which he will leave. What is the probability that the meeting actually takes place if each of them selects his moment of arrival at random during the interval from 12 noon to 1 PM?

96. A rod is broken into three pieces; the two break points are chosen at random. What is the probability that the three pieces can be joined at the ends to form a triangle?

97.* A rod of length L is broken at two points chosen at random. What is the probability that none of the three pieces has length greater than a given number a?

98. Three points A, B, and C on the circumference of a circle are chosen at random. What is the probability that all three angles of the triangle ABC are acute?

99.* A piece is broken off from each of three identical rods; the break points are chosen at random. What is the probability that a triangle can be formed from the three pieces obtained?

100.* A piece is broken off at random from each of three identical rods. What is the probability that an acute-angled triangle can be formed from the three pieces?

With this we close the set of problems illustrating the computation of probabilities. The problems presented here really belong to the "prehistory" of probability theory; the solution of such problems is connected with the origin of this branch of mathematics in the Seventeenth Century in the works of Pascal, Fermat, and Huygens. The further development of probability theory in the Eighteenth and early Nineteenth Centuries is connected with the names of Jacob Bernoulli, Laplace, and Gauss; in the works of these mathematicians many probability problems are collected and the methods of applying this new discipline to problems of science and engineering are considered for the first time. However, the final development of probability theory into an independent and deep science of great practical importance and with its own distinctive methods of investigation did not occur until the second half of the Nineteenth Century and the beginning of the Twentieth Century. The growth of probability theory continues at the present time.[10]

[10] *Supplementary reading:* E. B. Dynkin and V. A. Uspenskii, *Mathematicheskie Besedy* ("Mathematical Conversations"), section III, Moscow and Leningrad, Gostekhizdat, 1952; section III of this book is also available in German translation under the title "Mathematische Unterhaltungen" (Deutscher Verlag der Wissenschaften, East Berlin, 1956); John Riordan, *An Introduction to Combinatorial Analysis*, New York, Wiley, 1958.

SOLUTIONS

SOLUTIONS

I. INTRODUCTORY PROBLEMS

1. Let A, B, C be the three given points, and suppose l is a line equidistant from them. If A, B, C were all on the same side of l, they would lie on a line parallel to l, contradicting the hypothesis. Therefore two of the points are on one side of l and the third point is on the other side of l. Suppose for example that A and B are on one side of l, while C is on the other side. Then l must be parallel to the line AB and must pass through the midpoint of the perpendicular CP from P to the line AB (fig. 12).

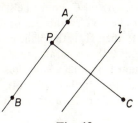

Fig. 12

These conditions completely determine l; conversely the line l so determined is actually equidistant from A, B, C. Hence there is a total of three lines equidistant from A, B, C (one separating each of the points from the other two points).

2. Let A, B, C, D be the four given points, and suppose Π is a plane which is equidistant from them. If A, B, C, D were all on the same side of Π, they would lie in a plane parallel to Π, contradicting the conditions of the problem. Consequently, only the following two cases are possible: (1) Three of the points lie on one side of Π and the fourth point is on the other side. (2) There are two of the points on each side of Π.

Consider case 1. Let A, B, C lie on one side of Π and D on the other side (fig. 13). The points A, B, C cannot be collinear, since if they were, all four points would be coplanar. Therefore A, B, C determine a unique plane, which must be parallel to Π. Moreover, Π must pass through the

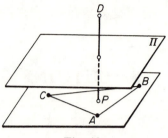

Fig. 13

midpoint of the perpendicular *DP* which joins *D* to the plane *ABC*. Thus there is one and only one plane Π equidistant from *A*, *B*, *C*, *D* with *A*, *B*, *C* on one side of it and *D* on the other side.

By the same reasoning there is exactly one plane equidistant from *A*, *B*, *C*, *D* with *C* (or *B* or *A*) on one side of it and the other three points on the other side of it. Thus there are a total of four planes in case 1.

Consider case 2. Let *A*, *B* lie on one side of Π and *C*, *D* on the other side (fig. 14). Since Π is equidistant from *A* and *B*, it must be parallel to the line *AB*. Likewise Π must be parallel to the line *CD*. Since *A*, *B*, *C*, *D* are not coplanar, the lines *AB* and *CD* must be skew. Now draw a plane Π₁ containing *AB* and parallel to *CD* (this can be done by drawing a line *l* through *A* parallel to *CD*; then Π₁ is the plane containing *l* and *B*). Also draw a plane Π₂ containing *CD* and parallel to *AB*. Then Π is parallel to these planes and equidistant from them, so it must pass through the midpoint of any perpendicular joining them (fig. 14). Thus there is one and only one plane Π equidistant from *A*, *B*, *C*, *D* with *A*, *B* on one side and *C*, *D* on the other.

The same reasoning shows that there is exactly one plane equidistant from *A*, *B*, *C*, *D* with *A*, *C* on one side and *B*, *D* on the other; and exactly

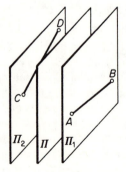

Fig. 14

one with A, D on one side and B, C on the other. Thus there are a total of three planes in case 2.

Combining the two cases gives a total of $4 + 3 = 7$ equidistant planes. If one considers the tetrahedron (triangular pyramid) with vertices A, B, C, D, then four of these seven planes are parallel to the faces of the tetrahedron and will pass through the midpoints of the corresponding altitudes (fig. 15a), and the other three planes are each parallel to a pair of opposite edges and equidistant from them (fig. 15b).

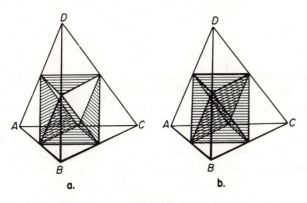

a.   b.

Fig. 15

3. Let A, B, C, D be the given points, and suppose s is a circle or straight line equidistant from them. Then A, B, C, D cannot all be on the same side of s. (By the two sides of a circle we mean, of course, the inside and the outside.) For they would then lie on a circle concentric with s or a line parallel to s according as s is a circle or a straight line. Hence there are two possibilities: (1) Three of the points lie on one side of s and the fourth point on the other side; (2) Two of the points lie on one side of s and the other two points on the other side.

Consider first case 1, and suppose for example that A, B, C are on one side of s, while D is on the other side. There is a unique circle or straight line t passing through A, B, C (fig. 16). Moreover t does not pass through D, by hypothesis. To be equidistant from the four points, s must be concentric with or parallel to t (according as t is a circle or a straight line) and must pass through the midpoint of the perpendicular DP from D to t. These conditions determine s uniquely; conversely the circle or straight line s so determined is indeed equidistant from A, B, C, D.

The same reasoning shows that there is exactly one circle or straight line equidistant from A, B, C, D with C (or B or A) on one side and the other three points on the other side. Thus there are a total of four circles or straight lines in case 1.

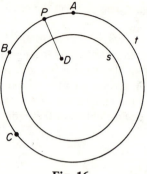

Fig. 16

Next consider case 2, and suppose for example that A, B are on one side of s, while C, D are on the other side (fig. 17a). If s is a circle, then its center O must be equidistant from A, B, and hence must lie on the perpendicular bisector p of AB. Similarly O must lie on the perpendicular bisector q of CD. Consequently, if p and q intersect, then O must be the point of intersection. Moreover the radius of s must be $\frac{1}{2}(OA + OC)$, since s must pass midway between the circles s_1 and s_2 which have O as center and OA, OC respectively as radii. (In this case there is no straight line equidistant from A, B, C, D, since any such line would be parallel to AB and CD, thus making p parallel to q).

If p and q are parallel (fig. 17b), then AB is parallel to CD. In this case the only possibility for s is a straight line parallel to AB and lying midway between AB and CD.

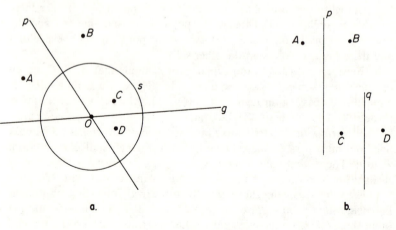

a.   b.

Fig. 17

The case in which p coincides with q cannot arise, since then there would be a circle passing through A, B, C, D (see fig. 18; the perpendicular bisector r of AC intersects p in a point E equidistant from A, B, C, D).

Thus in any case there is exactly one circle or straight line which is equidistant from A, B, C, D with A, B on one side of it, and C, D on the other. By the same reasoning there is exactly one such circle separating A, C from B, D and one separating A, D from B, C. Hence there are altogether 3 circles or straight lines in case 2.

Combining 1 and 2, we get a total of $4 + 3 = 7$ circles or straight lines equidistant from the four points.

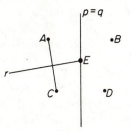

Fig. 18

4. Let A, B, C, D, E be the given points, and suppose Σ is a sphere or plane equidistant from them. Then A, B, C, D, E cannot all be on the same side of Σ. (By the two sides of a sphere we mean the inside and the outside.) For if they were, they would lie on a sphere concentric with Σ or on a plane parallel to Σ, according as Σ is a sphere or a plane. Consequently there are two possible cases: (1) Four of the points lie on one side of Σ and the fifth point on the other side; (2) three of the points lie on one side of Σ and the other two on the other side.

Consider first case 1, and suppose that A, B, C, D are on one side of Σ, while E is on the other side. Then A, B, C, D do not all lie on a circle or a straight line, since if they did, A, B, C, D, E would be cospherical (i.e., all on a sphere) or coplanar. Consequently, there is a unique sphere or plane S passing through A, B, C, D. If S is a sphere, then Σ is a sphere concentric with S (fig. 19). Moreover Σ must pass through the midpoint of ME, where O is the center of S, and M is the point where OE meets S. If S is a plane, then Σ is a plane parallel to S (fig. 20), and Σ must pass through the midpoint of EP, the perpendicular from E to S. Thus, in either case there is exactly one sphere or plane Σ equidistant from A, B, C, D, E and such that A, B, C, D are on one side of it and E is on the other side.

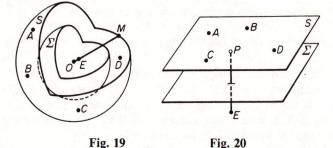

Fig. 19   **Fig. 20**

The same reasoning shows that there is exactly one sphere or plane equidistant from A, B, C, D, E with D (or C or B or A) on one side and the other four points on the other side. Thus there are a total of five spheres or planes in case 1.

Next consider case 2, and suppose A, B, C are on one side of Σ, while D, E are on the other side. There is exactly one straight line or circle s which passes through A, B, C. Suppose first that s is a circle, and denote its center by F (fig. 21). If Σ is a sphere, then its center O must be equidistant from A, B, and C; consequently O must lie on the line p which is perpendicular to the plane ABC and passes through F. On the other hand, O must be equidistant from D and E, and therefore must lie in the plane Π which passes through the midpoint of the segment DE and is perpendicular to it. Consequently, if p intersects Π, then O must be the point of intersection. Moreover the radius of Σ must be equal to $\frac{1}{2}(OA + OD)$, since Σ must pass midway between the spheres Σ_1 and Σ_2 which have O as center and OA, OD respectively as radii.

If p and Π do not meet, i.e., if they are parallel (fig. 22), then DE is parallel to the plane ABC. In this case the only plane or sphere equidistant from A, B, C, D, E with A, B, C on one side and D, E on the other is the

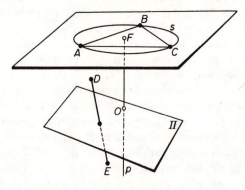

Fig. 21

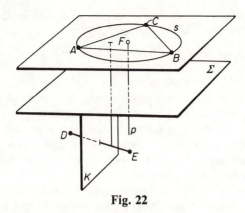

Fig. 22

plane parallel to ABC and passing through the midpoint of the perpendicular to ABC from any point of DE. The case in which p lies in Π cannot arise, since then there would be a sphere passing through all five points.

Now suppose that s is a straight line (fig. 23). In this case Σ must be a plane, since three collinear points on the same side of a sphere cannot be equidistant from that sphere. As in the solution to problem 3, there is a unique plane Σ which passes midway between the (unique) pair of parallel planes containing the skew lines ABC and DE; this plane is the desired one.

Thus in any case there is exactly one sphere or plane which is equidistant from A, B, C, D, E with A, B, C on one side of it and D, E on the other. By the same reasoning, for every pair of points selected from A, B, C, D, E, there is exactly one sphere or plane Σ equidistant from A, B, C, D, E with the selected pair on one side of Σ and the remaining three points on the other side. There are $\binom{5}{2} = 10$ such pairs (AB, AC, AD, AE, BC, BD, BE, CD, CE, DE), and so there are altogether 10 spheres or planes in case 2.

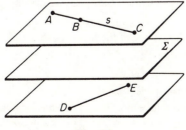

Fig. 23

Combining cases 1 and 2 we get a total of $5 + 10 = 15$ spheres and planes equidistant from the five points.

5. The problem amounts to determining how many points (the centers of the required spheres) are equidistant from the four faces of the tetrahedron.

The locus of all points equidistant from the two faces of a dihedral angle is the bisector of the dihedral angle, i.e., the plane which divides it into two congruent dihedral angles. Since any two intersecting planes form two pairs of vertical dihedral angles, the locus of all points equidistant from two intersecting planes consists of two planes which pass through the line in which the given planes intersect, namely, the bisectors of the dihedral angles formed by the given planes.

The locus of all points equidistant from the faces of a trihedral angle is the line where the bisectors of the dihedral angles of which it is composed meet; this line passes through the vertex of the trihedral angle.

Since three planes intersecting in a point P form four pairs of vertical trihedral angles, the locus of all points equidistant from the given planes consists of four straight lines through P.

Now let $\Pi_1, \Pi_2, \Pi_3, \Pi_4$ be the planes of the faces of the tetrahedron T. The above remarks show that the locus of all points equidistant from Π_1, Π_2, and Π_3 consists of four straight lines $\beta_1, \beta_2, \beta_3, \beta_4$. Furthermore, the locus of all points equidistant from Π_1 and Π_4 consists of two planes B_1 and B_2. Every point at which one of the lines $\beta_1, \beta_2, \beta_3, \beta_4$ meets one of the planes B_1, B_2 is equidistant from $\Pi_1, \Pi_2, \Pi_3, \Pi_4$ (and these are the only such points). Hence there can be at most eight points equidistant from $\Pi_1, \Pi_2, \Pi_3, \Pi_4$ (there may be less, since one of the β's might be parallel to one of the B's). This means that there are at most eight spheres tangent to $\Pi_1, \Pi_2, \Pi_3, \Pi_4$. Of these eight spheres, one lies inside the tetrahedron (the inscribed sphere, fig. 24a), four lie outside the tetrahedron, inscribed in each of the four trihedral angles shown in fig. 24b, and the other three are each inscribed in an interior dihedral angle and the opposite exterior dihedral angle (fig. 24c).

In special cases a tetrahedron can have less than eight spheres tangent to the planes of its faces. For example, a regular tetrahedron has only five such spheres (only those of the kinds shown in figs. 24a and 24b).

Remark. Let A_1, A_2, A_3, A_4 be the areas of the four faces of the tetrahedron T. It can be shown that if one of the three equations

(1) $\quad A_1 + A_2 = A_3 + A_4$

(2) $\quad A_1 + A_3 = A_2 + A_4$

(3) $\quad A_1 + A_4 = A_2 + A_3$

holds, then there is no sphere inscribed in the interior dihedral angle formed by the planes on one side of the equation and the exterior dihedral angle formed by those on the other side. Conversely if the equation does not hold, then there

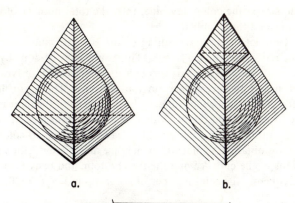

a.   b.

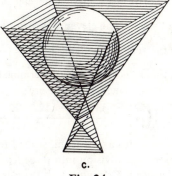

c.

Fig. 24

will be such a sphere. Therefore:

    1. If none of the equations (1), (2), or (3) holds, there are eight spheres tangent to all the faces.

    2. If one of the equations holds there are only seven tangent spheres.

    3. If two of the equations hold, there are only six tangent spheres. In this case the faces are equal in pairs; for example, if (2) and (3) hold, then $A_1 = A_2$ and $A_3 = A_4$.

    4. If (1), (2), and (3) all hold (in which case $A_1 = A_2 = A_3 = A_4$), there are only five tangent spheres.[1]

6. Suppose that the six colors are yellow, blue, red, green, white, and black. Turn the cube so that the green face is on the bottom. The top face can then be any of the other five colors. Furthermore, no two colorings in which the top face is painted different colors (the bottom face being green in each case) can be obtained from one another by rotating the cube. So if we can determine the number of colorings in which the top face is painted some specific color, say blue, then the total number of colorings will be five times this number.

Select one of the four remaining colors, say red, and turn the cube so that the red face is the back face. This can always be done by rotating the cube about the vertical axis through its center (which leaves the blue face on top and the green face on the bottom). There are now three colors left, yellow, black, and white, with which to color the three remaining faces (the front face and the two side faces). The front face can therefore be colored in three ways, each of which leaves two ways to color the side faces. So we obtain six colorings which are all different, since there are no rotations of the cube leaving the top, bottom, and back faces fixed.

Consequently the total number of colorings is $5 \times 6 = 30$.

7. The eleven boys who form the first team can be chosen in $\binom{33}{11} = \dfrac{33 \cdot 32 \cdots 23}{11!}$ different ways. The eleven boys who form the second team can be chosen from the remaining 22 in $\binom{22}{11} = \dfrac{22 \cdot 21 \cdots 12}{11!}$ different ways. Then the third team is completely determined, since it consists of the eleven remaining boys. Therefore there are

$$\binom{33}{11}\binom{22}{11} = \frac{33 \cdot 32 \cdots 12}{(11!)^2} = \frac{33!}{(11!)^3}$$

ways of dividing the boys into a first team, second team, and third team. But in this enumeration each division into three teams is counted $3! = 6$ times, once for each way of deciding which team is to be called "first," "second," and "third." Thus the number of divisions into three teams is

$$\frac{33!}{3!\,(11!)^3} = 22,754,499,243,840.$$

Remark. In exactly the same way one can show that the number of ways of dividing nk objects into k sets of n objects is $\dfrac{(nk)!}{(n!)^k k!}$.

8. Let the 11 flavors be $F_1, F_2, \ldots, F_{11}$. Then any choice of 6 ice cream cones can be described by listing the flavors in increasing order. For example, $F_1 F_1 F_3 F_5 F_6 F_8$ would mean a choice of 2 ice cream cones of flavor F_1, one of F_3, etc. Now let us draw vertical lines to separate the F_1's from the F_2's, the F_2's from the F_3's, $\ldots$, and finally the F_{10}'s from the F_{11}'s. In the above example we would get $F_1 F_1 | \; | F_3 | F_5 \; | F_6 | F_8 \; | \; |$. The result is a sequence of 16 symbols of which 6 are F's and 10 are vertical lines. Conversely, start with a row of 16 dots:

$$\cdot \; \cdot \; \cdot \; \cdot \; \cdot \; \cdot \; \cdot \; \cdot \; \cdot \; \cdot \; \cdot \; \cdot \; \cdot \; \cdot \; \cdot \; \cdot$$

Pick any 10 of the dots and draw vertical lines through them. For example, we might get

$$| \; \cdot \; | \; | \; \cdot \; | \; \cdot \; \cdot \; | \; | \; | \; \cdot \; | \; | \; | \; \cdot$$

Then this scheme determines a choice of 6 ice cream cones in which the dots to the left of the first vertical line are F_1's, the dots between the first and second vertical lines are F_2's, etc. In the above example we would get $F_2F_4F_5F_5F_8F_{11}$.

It follows that the number of ways of choosing the 6 ice cream cones is equal to the number of ways of choosing 10 dots out of 16. This number is $\binom{16}{10} = 8008$.

Remark. If there are n different flavors available and the number of ice cream cones to be selected is m, then the number of ways of choosing them is $\binom{m+n-1}{n-1} = \binom{m+n-1}{m}$. This can be proved in exactly the same way as the special case $n = 11$, $m = 6$ treated above.

9. Given any 5 members of the group, there must be a lock for which none of them has the key. But each of the other 6 members must have this key, since the addition of one more member to the 5 forms a majority. Therefore the number of locks must be at least equal to the number of ways in which 5 people can be selected out of the 11 scientists, i.e., $\binom{11}{5} = 462$.

Now let A be one of the scientists. We have just seen that given any set of 5 scientists selected from the remaining 10, there is a lock which they cannot open, and that A has a key to this lock. So A has at least $\binom{10}{5} = 252$ keys.

The lower bounds just derived can actually be attained by using 462 locks, one for each set S of 5 scientists. If L is the lock associated with the set S, then only the other 6 scientists are given keys to L. A majority M can always open the safe, since for any lock L there are only 5 people not having keys to L, so that some one in M has a key.

Remark. If there are n scientists and if it is required that the safe can be opened if and only if at least m of them are present, then the minimum number of locks is $\binom{n}{m-1}$, and the minimum number of keys held by each scientist is $\binom{n-1}{m-1}$. This indicates that such a system of safekeeping is impractical, since even for comparatively small values of m and n, opening the safe would require a whole day of fumbling with keys.

10. After the first time around, all numbers which leave a remainder of 1 upon division by 15 will have been marked; the last of these will be 991. The first number to be marked on the second time around will be $991 + 15 - 1000 = 6$; further, on the second time around, all numbers which leave a remainder of 6 upon division by 15 will be marked (the last of them will be 996). The first number to be marked on the third time around will be $996 + 15 - 1000 = 11$; further, on the third time around, all numbers which leave a remainder of 11 upon division by 15 will be marked (the last of them will be 986). The first number to be marked on the fourth time around would be $986 + 15 - 1000 = 1$. Since this number has already been marked earlier, by continuing to count in 15's around the circle, we would keep hitting numbers already marked.

Thus, those numbers and only those numbers which leave remainders of 1, 6, or 11 upon division by 15 will get marked. But these are precisely the numbers which leave a remainder of 1 upon division by 5. There are $1000/5 = 200$ such numbers among the first 1000 (namely, 1, 6, 11, 16, ..., $996 = 5 \cdot 199 + 1$). Consequently $1000 - 200 = 800$ numbers remain unmarked.

11a. *First solution.* Let us first compute how many of the integers under consideration *do not have any* 1's among their digits. By adjoining 0 and deleting 10,000,000,000, we obtain a sequence of 10^{10} numbers in which there is one more number with no 1's among its digits than there was before. Let us agree to write before each number of less than ten digits enough zeros to raise the number of digits to ten. The new sequence of numbers consists of 10^{10} integers, beginning with 0,000,000,000 and ending with 9,999,999,999. If one of these numbers has no 1's among its digits, then its first digit must be one of the nine numbers 0, 2, 3, 4, 5, 6, 7, 8, 9. The second digit will likewise have to be one of these nine numbers. By pairing each of the nine possible values for the first digit with each of the nine possible values for the second, we obtain a total of 9^2 different possibilities for the first pair of digits. In exactly the same way, we obtain 9^3 possibilities for the first three digits of our number, 9^4 different possibilities for the first four digits, etc., and finally 9^{10} different possibilities for the first ten digits. But this means that among the integers from 0,000,000,000 to 9,999,999,999 there are 9^{10} different numbers which have no 1's among their digits. Consequently, there are exactly

$$9^{10} - 1 = 3,486,784,401 - 1 = 3,486,784,400$$

such numbers among the integers from 1 to 10,000,000,000. Therefore, there are

$$10,000,000,000 - 3,486,784,400 = 6,513,215,600$$

of them in which 1's occur among the digits.

Second solution. Let us compute how many of the integers under consideration *have* 1's among their digits. By a *ten-row* we mean a set of ten integers which consists of some multiple of ten and the next nine integers after it (for example, $\{210, 211, 212, \ldots, 219\}$); by a *hundred-row* is meant a set of 100 integers which consists of a multiple of 100 and the next 99 integers after it (for example, $\{32400, 32401, 32402, \ldots, 32499\}$); by a *thousand-row* we mean a set of 1000 integers which consists of some multiple of 1000 and the next 999 integers after it, etc. Note that 0, being a multiple of any number, is a multiple of 10, of 100, of 1000, etc.; hence $\{0, 1, \ldots, 9\}$, $\{0, 1, 2, \ldots, 99\}$, etc., count respectively as the first ten-row, the first hundred-row, etc.

In the first ten-row there is exactly one number among whose digits a 1 occurs, namely the number 1. In the first hundred-row, each ten-row except the second (which consists of the numbers from 10 to 19) will contain exactly one number with a 1 among its digits; the second ten-row consists entirely of numbers with 1's among their digits.

Consequently, the first hundred-row contains

$$10 + 9 \cdot 1$$

numbers which have 1's among their digits: ten in the second ten-row and one in each of the other 9 ten-rows.

In the first thousand-row, each hundred-row except the second will therefore contain exactly $10 + 9 \cdot 1$ integers which have 1's among their digits; the second hundred-row consists entirely of numbers with 1's among their digits. Consequently, the first thousand-row contains

$$100 + 9(10 + 9 \cdot 1) = 10^2 + 9 \cdot 10 + 9^2$$

numbers which have 1's among their digits.

In the first ten-thousand row, each thousand-row except the second will contain exactly $10^2 + 9 \cdot 10 + 9^2$ numbers which have 1's among their digits; the second thousand-row will consist entirely of such numbers. Thus among the first 10,000 nonnegative integers there will be

$$1000 + 9(10^2 + 9 \cdot 10 + 9^2) = 10^3 + 9 \cdot 10^2 + 9^2 \cdot 10 + 9^3$$

numbers which have 1's among their digits.

Continuing to reason in this fashion, we see that among the first 10,000,000,000 nonnegative integers (the numbers from 0 to 9,999,999,999) there are

$$\begin{aligned}
10^9 + 9 \cdot 10^8 &+ 9^2 \cdot 10^7 + \cdots + 9^8 \cdot 10 + 9^9 \\
&= 10^9(1 + \tfrac{9}{10} + (\tfrac{9}{10})^2 + \cdots + (\tfrac{9}{10})^9) \\
&= 10^9 \frac{1 - (9/10)^{10}}{1 - 9/10} \\
&= \frac{10^9 - 9^{10}/10}{1/10} = 10^{10} - 9^{10}
\end{aligned}$$

numbers which have 1's among their digits. Among the numbers from 1 to 10,000,000,000, the number of integers with 1's among their digits will be one greater, namely

$$10^{10} - 9^{10} + 1 = 6{,}513{,}215{,}600.$$

This result of course coincides with that obtained in the first solution. It shows that among the numbers under consideration there are more with 1's among their digits than without.

11b. Among the integers from 1 to 222,222,222 there are 22,222,222 ending in a 0 (namely, the numbers 10, 20, 30, . . . , 222,222,220). In order to determine how many integers have a 0 in the next to last position, notice that what comes before this 0 can be anything from 1 to 2,222,222, while what comes after it can be anything from 0 to 9. Therefore there are $2,222,222 \cdot 10 = 22,222,220$ integers with a 0 as their next to last digit. Similarly there are $222,222 \cdot 100 = 2,222,200$ integers with a 0 as their second-from-last digit, because what comes before this 0 can be anything from 1 to 222,222, while what comes after it can be anything from 00 to 99. Continuing in this way we see that the total number of 0's is

$$22,222,222 + 22,222,220 + 22,222,200 + 22,222,000 + 22,220,000$$
$$+ \; 22,200,000 + 22,000,000 + 20,000,000 = 175,308,642.$$

II. THE REPRESENTATION OF INTEGERS AS SUMS AND PRODUCTS

12a. The expression $\#(A) + \#(B)$ counts the number of elements in A or B, but counts the elements of $A \cap B$ twice. So by subtracting $\#(A \cap B)$ we get exactly the number of elements in $A \cup B$.

12b. Consider an element e which is in only one of the three sets, say in A. Then in the expression

$$\#(A) + \#(B) + \#(C) - \#(A \cap B)$$
$$- \#(A \cap C) - \#(B \cap C) + \#(A \cap B \cap C), \quad (1)$$

e is counted exactly once, namely in the term $\#(A)$. Next consider an element f which is in exactly two of the sets, say A and B. Then f is counted positively in the terms $\#(A)$ and $\#(B)$, and negatively in the term $-\#(A \cap B)$. Hence it is counted a net of $1 + 1 - 1 = 1$ time in the expression (1). Finally suppose g is an element in all three of the sets A, B, C. Then g is counted by each term of (1), and is therefore counted a net of $1 + 1 + 1 - 1 - 1 - 1 + 1 = 1$ time. This analysis shows that expression (1) counts each element of $A \cup B \cup C$ once. On the other hand, elements not in $A \cup B \cup C$ are not counted in any of its terms, and therefore (1) is equal to $\#(A \cup B \cup C)$.

12c. The general case can be treated by the same reasoning as that used in part b. We must show that in the expression

$$\#(A_1) + \#(A_2) + \cdots + \#(A_m) - \#(A_1 \cap A_2) - \#(A_1 \cap A_3) - \cdots$$
$$- \#(A_{m-1} \cap A_m) + \#(A_1 \cap A_2 \cap A_3) + \cdots$$
$$+ (-1)^{m-1}\#(A_1 \cap A_2 \cap \cdots \cap A_m), \quad (2)$$

every element of $A_1 \cup \cdots \cup A_m$ is counted with a net coefficient of 1. Note that elements not in $A_1 \cup \cdots \cup A_m$ are not counted at all by (2). Suppose an element e is in exactly $h \geq 1$ of the sets $A_1, \ldots, A_m$. For definiteness, suppose it is in $A_1, \ldots, A_h$, but not in $A_{h+1}, \ldots, A_m$. Then e will be counted in each of the h terms $\#(A_1), \ldots, \#(A_h)$ of (2). It will be counted negatively in $\binom{h}{2}$ of the terms $-\#(A_i \cap A_j)$ (namely, the terms where $1 \leq i \leq j \leq h$; there are $\binom{h}{2}$ such terms because there are $\binom{h}{2}$ ways of choosing the integers i, j from among the numbers $1, \ldots, h$). Similarly e will be counted by $\binom{h}{3}$ of the terms $\#(A_i \cap A_j \cap A_k)$, etc. The total number of times e is counted is therefore

$$\binom{h}{1} - \binom{h}{2} + \binom{h}{3} - \cdots + (-1)^{h-1}\binom{h}{h}.$$

We must show that this expression is equal to 1. To see this we use the binomial theorem

$$(a + b)^h = \binom{h}{0}a^h + \binom{h}{1}a^{h-1}b + \binom{h}{2}a^{h-2}b^2 + \cdots + \binom{h}{h}b^h$$

Putting $a = 1$, $b = -1$, the left-hand side vanishes, so we get

$$0 = \binom{h}{0} - \binom{h}{1} + \binom{h}{2} - \cdots + (-1)^h\binom{h}{h}.$$

Transposing all terms but $\binom{h}{0}$ to the left-hand side, and using the fact that $\binom{h}{0} = 1$, we see that

$$\binom{h}{1} - \binom{h}{2} + \binom{h}{3} - \cdots + (-1)^h\binom{h}{h} = 1,$$

which completes the proof.

13a. We will solve this problem by applying the principle of inclusion and exclusion (see problem 12). Let A be the set of all positive integers ≤ 999 which are multiples of 5, and let B consist of all multiples of 7 in the same range. Then $\#(A) = [999/5] = 199$, since A consists of the numbers $5, 10, 15, \ldots, 995 = 199 \cdot 5$. Similarly $\#(B) = [999/7] = 142$, since B consists of the numbers $7, 14, 21, \ldots, 994 = 142 \cdot 7$. Now $A \cap B$ consists of all positive integers ≤ 999 which are multiples of 5 and 7, that is, the multiples of 35. The same reasoning as above then shows that $\#(A \cap B) = [999/35] = 28$. By the principle of inclusion and exclusion, $\#(A \cup B) = 199 + 142 - 28 = 313$. The elements of $A \cup B$ are the positive integers < 1000 which are divisible by either 5 or 7. Since we want to know how many integers in this range are *not* divisible by either

5 or 7, we must subtract 313 from 999, getting $999 - 313 = 686$ as the final answer.

13b. Let A and B be the same sets as in part a, and let C be the set of all positive integers ≤ 999 which are multiples of 3. Then $A \cap C$ consists of the multiples of $5 \cdot 3 = 15$, $B \cap C$ of the multiples of $7 \cdot 3 = 21$, and $A \cap B \cap C$ of the multiples of $5 \cdot 7 \cdot 3 = 105$ in this range. Reasoning as in part a we see that

$$\#(C) = \left[\frac{999}{3}\right] = 333, \qquad \#(A \cap C) = \left[\frac{999}{15}\right] = 66,$$

$$\#(B \cap C) = \left[\frac{999}{21}\right], \quad \text{and} \quad \#(A \cap B \cap C) = \left[\frac{999}{105}\right] \cdot 9.$$

By the principle of inclusion and exclusion,

$$\#(A \cup B \cup C) = 199 + 142 + 333 - 28 - 66 - 47 + 9 = 542.$$

The elements $A \cup B \cup C$ are the positive integers ≤ 999 which are divisible by either 5 or 7 or 3. Since we want to know how many integers in this range are *not* divisible by 5, 7, or 3 we must subtract 542 from 999, getting $999 - 542 = 457$ as the solution to the problem.

Remark. By applying the general case of the principle of inclusion and exclusion, one can obtain the following extension of problem 13. Let positive integers $p_1, p_2, \ldots, p_m$ be given, no two of which have a common factor > 1 (in the terminology of number theory, p_i and p_j are relatively prime when $i \neq j$). Let N be a positive integer. Then the number of positive integers $\leq N$ which are not divisible by any of the numbers $p_1, p_2, \ldots, p_m$, is

$$N - \left[\frac{N}{p_1}\right] - \cdots - \left[\frac{N}{p_m}\right] + \left[\frac{N}{p_1 p_2}\right] + \left[\frac{N}{p_1 p_3}\right] + \cdots + \left[\frac{N}{p_{m-1} p_m}\right]$$
$$- \left[\frac{N}{p_1 p_2 p_3}\right] - \cdots + (-1)^m \left[\frac{N}{p_1 p_2 \cdots p_m}\right].$$

Problem 13b is the case where $N = 999$, $m = 3$, $p_1 = 3$, $p_2 = 5$, $p_3 = 7$.

14. This problem is closely related to the preceding one. The factorization of 1260 into prime factors is $1260 = 2^2 \cdot 3^2 \cdot 5 \cdot 7$. Consequently the problem amounts to determining the number of positive integers ≤ 1260 which are not divisible by 2, 3, 5, or 7. Let A consist of the positive integers ≤ 1260 which are multiples of 2; let B consist of the multiples of 3 in the same range, C of the multiples of 5, and D of the multiples of 7. By the principle of inclusion and exclusion,

$$\begin{aligned}
\#(A \cup B \cup C \cup D) = {} & \#(A) + \#(B) + \#(C) + \#(D) \\
& - \#(A \cap B) - \#(A \cap C) - \#(A \cap D) \\
& - \#(B \cap C) - \#(B \cap D) - \#(C \cap D) \\
& + \#(A \cap B \cap C) + \#(A \cap B \cap D) + \#(A \cap C \cap D) \\
& + \#(B \cap C \cap D) - \#(A \cap B \cap C \cap D).
\end{aligned}$$

Now the reasoning of problem 13 shows that

$$\#(A) = \frac{1260}{2} = 630$$

$$\#(B) = \frac{1260}{3} = 420$$

$$\#(C) = \frac{1260}{5} = 252$$

$$\#(D) = \frac{1260}{7} = 180$$

$$\#(A \cap B) = \frac{1260}{2 \cdot 3} = 210$$

$$\#(A \cap C) = \frac{1260}{2 \cdot 5} = 126$$

$$\#(A \cap D) = \frac{1260}{2 \cdot 7} = 90$$

$$\#(B \cap C) = \frac{1260}{3 \cdot 5} = 84$$

$$\#(B \cap D) = \frac{1260}{3 \cdot 7} = 60$$

$$\#(C \cap D) = \frac{1260}{5 \cdot 7} = 36$$

$$\#(A \cap B \cap C) = \frac{1260}{2 \cdot 3 \cdot 5} = 42$$

$$\#(A \cap B \cap D) = \frac{1260}{2 \cdot 3 \cdot 7} = 30$$

$$\#(A \cap C \cap D) = \frac{1260}{2 \cdot 5 \cdot 7} = 18$$

$$\#(B \cap C \cap D) = \frac{1260}{3 \cdot 5 \cdot 7} = 12$$

$$\#(A \cap B \cap C \cap D) = \frac{1260}{2 \cdot 3 \cdot 5 \cdot 7} = 6.$$

Therefore our formula gives

$$\#(A \cup B \cup C \cup D) = 630 + 420 + 252 + 180 - 210 - 126 - 90$$
$$- 84 - 60 - 36 + 42 + 30 + 18 + 12 - 6 = 972.$$

Now $A \cup B \cup C \cup D$ consists of the positive integers ≤ 1260 which are divisible by at least one of the numbers 2, 3, 5, 7. To find out how many integers are *not* divisible by 2, 3, 5, or 7 we therefore subtract 972 from 1260, getting 288.

Remark. As in the remark to problem 13, the principle of inclusion and exclusion for m sets can be applied to prove the following generalization of the result just obtained: If N is a positive integer, and $N = p_1^{a_1} p_2^{a_2} \cdots p_m^{a_m}$ is its factorization into primes, then $\varphi(N)$, the number of positive integers $\leq N$ and relatively prime to N, is given by the formula

$$\varphi(N) = N - \frac{N}{p_1} - \frac{N}{p_2} - \cdots - \frac{N}{p_m} + \frac{N}{p_1 p_2} + \frac{N}{p_1 p_3} + \cdots + \frac{N}{p_{m-1} p_m} - \frac{N}{p_1 p_2 p_3}$$

$$- \cdots - \frac{N}{p_{m-2} p_{m-1} p_m} + \cdots + (-1)^m \frac{N}{p_1 p_2 \cdots p_m}$$

$$= N \left(1 - \frac{1}{p_1}\right) \left(1 - \frac{1}{p_2}\right) \cdots \left(1 - \frac{1}{p_m}\right).$$

This last expression greatly simplifies the calculation of $\varphi(N)$. Thus in the above example we get

$$\varphi(1260) = 1260(1 - \tfrac{1}{2})(1 - \tfrac{1}{3})(1 - \tfrac{1}{5})(1 - \tfrac{1}{7}) = 1260 \cdot \tfrac{1}{2} \cdot \tfrac{2}{3} \cdot \tfrac{4}{5} \cdot \tfrac{6}{7} = 288.$$

15. We will first determine the number of integers x in the range $1 \leq x \leq 10,000$ such that $2^x - x^2$ is divisible by 7. Subtracting this number from 10,000 then gives the answer to the problem.

Now $2^x - x^2$ is divisible by 7 if and only if 2^x and x^2 both leave the same remainder when divided by 7. So it is natural to study these remainders. The first few powers of 2 are 2, 4, 8, 16, 32, 64, ..., and their remainders when divided by 7 are 2, 4, 1, 2, 4, 1, ... These remainders will keep repeating with a period of 3, so that 2, 4, and 1 are the only remainders which 2^x can have. To prove that 2^x and 2^{x+3} have the same remainder when divided by 7, note that $2^{x+3} - 2^x = 8 \cdot 2^x - 2^x = 7 \cdot 2^x$ is a multiple of 7.

Next we will make a similar analysis of the remainders of x^2 when divided by 7. Putting $x = 1, 2, 3, 4, 5, 6, 7$ we have $x^2 = 1, 4, 9, 16, 25, 36, 49$, so that the remainders are 1, 4, 2, 2, 4, 1, 0. Thereafter the remainders will repeat with a period of 7, that is $(x + 7)^2$ will have the same remainder as x^2. This is because $(x + 7)^2 - x^2 = 14x + 49 = 7(2x + 7)$ is a multiple of 7.

Combining these results, we see that the remainders of *both* 2^x and x^2 will repeat after a period of length $3 \cdot 7 = 21$. A tabulation of the first 21 value of x is shown in the chart on page 57.

There are 6 cases in this range where $2^x - x^2$ is divisible by 7 (namely $x = 2, 4, 5, 6, 10, 15$). By periodicity the next 21 values of x will give 6 more cases, and so on. Now $10,000 = 21 \cdot 476 + 4$, so the integers from 1 to 10,000 split into 476 groups of 21 and 4 extra numbers at the end.

x	1	2	3	4	5	6	7	8	9	10	11	12	13	14	15	16	17	18	19	20	21
2^x	2	4	1	2	4	1	2	4	1	2	4	1	2	4	1	2	4	1	2	4	1
x^2	1	4	2	2	4	1	0	1	4	2	2	4	1	0	1	4	2	2	4	1	0

The 476 groups contribute $476 \cdot 6 = 2856$ values of x, and the remaining four numbers contribute 2 values of x (namely $x = 9998$ and $x = 10{,}000$). So the total is 2858; subtracting this from 10,000 we get $10{,}000 - 2858 = 7142$ values of x for which $2^x - x^2$ is not divisible by 7.

16. If $x^2 + y^2$ is divisible by 49, then $x^2 + y^2$ is divisible by 7. But x^2 can only have 0, 1, 2, or 4 as its remainder upon division by 7. (See the solution to problem 15.) The same is true of y^2. Now it is easy to verify that of all sums of two of the numbers 0, 1, 2, 4, only the sum $0 + 0 = 0$ is divisible by 7. Consequently, $x^2 + y^2$ is divisible by 7 only when x^2 and y^2 are divisible by 7, that is, only when x and y are divisible by 7. Conversely, if x and y are two numbers divisible by 7, then the sum $x^2 + y^2$ is divisible by 49. Thus the number we are seeking is the number of different pairs of positive integers x and y which are both less than 1000 and divisible by 7.

Since $1000 = 7 \cdot 142 + 6$, there are 142 multiples of 7 between 1 and 1000. Pairing each of the 142 values for x with each of the 142 values for y, we obtain a total of 142^2 pairs (x,y). Of these, 142 pairs are of the form (x,x); each other pair occurs twice, once as (x,y) and once as (y,x). Consequently, the total number of different pairs (x,y) equals

$$\frac{142^2 - 142}{2} + 142 = \frac{142 \cdot 143}{2} = 10{,}153.$$

17. Since $1{,}000{,}000 = 2^6 \cdot 5^6$, each of its divisors has the form $2^a \cdot 5^b$, and a decomposition of 1,000,000 into a product of three factors has the form

$$1{,}000{,}000 = (2^{a_1} \cdot 5^{b_1})(2^{a_2} \cdot 5^{b_2})(2^{a_3} \cdot 5^{b_3});$$

here a_1, a_2, a_3, b_1, b_2 and b_3 are nonnegative integers which satisfy the conditions

$$a_1 + a_2 + a_3 = 6, \quad b_1 + b_2 + b_3 = 6.$$

Let us compute how many such systems of numbers a_1, a_2, a_3, b_1, b_2, b_3 there are.

If $a_1 = 6$, then a_2 and a_3 must equal 0; thus in this case we have only one possible system of numbers a_1, a_2, a_3.

If $a_1 = 5$, then two systems are possible:

$$a_1 = 5, \ a_2 = 1, \ a_3 = 0 \quad \text{and} \quad a_1 = 5, \ a_2 = 0, \ a_3 = 1.$$

If $a_1 = 4$, then three systems are possible:

$$a_1 = 4, a_2 = 2, a_3 = 0; \quad a_1 = 4, a_2 = 1, a_3 = 1;$$
$$a_1 = 4, a_2 = 0, a_3 = 2.$$

It can be shown in the same way that if $a_1 = 3$, then four systems are possible, if $a_1 = 2$, then five systems, if $a_1 = 1$, then six systems, and if $a_1 = 0$, then seven systems. Thus the total number of systems of non-negative integers a_1, a_2, a_3 which satisfy $a_1 + a_2 + a_3 = 6$ is

$$1 + 2 + 3 + 4 + 5 + 6 + 7 = 28.$$

By the same argument there are exactly 28 systems of numbers b_1, b_2, b_3 which satisfy $b_1 + b_2 + b_3 = 6$. Since any combination of such a triple of numbers a_1, a_2, a_3 with a triple b_1, b_2, b_3 yields a decomposition of 1,000,000 into a product of three factors, the total number of decompositions is

$$28 \cdot 28 = 784.$$

However, in this enumeration, factorizations which differ only in the order of the factors have been counted separately; that is, some factorizations are counted several times each. Let us determine how many times each factorization occurs.

(1) Exactly one factorization, namely

$$10^6 = (2^2 \cdot 5^2)(2^2 \cdot 5^2)(2^2 \cdot 5^2),$$

occurs only once.

(2) If in a decomposition of 10^6 into a product of three factors, two of the factors are equal (and the third factor different from them), then the factorization occurs three times: the unduplicated factor can come either first, second, or third.

Let us compute the number of such factorizations. Let the factor which is repeated in the factorization be $2^a \cdot 5^b$; that is, the factorization is to have the form $10^6 = (2^a \cdot 5^b) \cdot (2^a \cdot 5^b)(2^{6-2a} \cdot 5^{6-2b})$; consequently, a can take any of the values 0, 1, 2, 3 and b can likewise take any of the values 0, 1, 2, 3. Since combining any such a with any such b yields a factorization of the required form, the total number of possibilities is $4 \cdot 4 = 16$. One possibility, namely, $a = 2$, $b = 2$, must be discarded, since it leads to the factorization

$$10^6 = (2^2 \cdot 5^2) \cdot (2^2 \cdot 5^2) \cdot (2^2 \cdot 5^2)$$

treated above. Thus, 15 factorizations occur three times each.

(3) The remaining factorizations occur six times each. In fact, if no two of the three factors are equal, then the following six orders for the factors are possible:

$$(2^{a_1} \cdot 5^{b_1}) \cdot (2^{a_2} \cdot 5^{b_2}) \cdot (2^{a_3} \cdot 5^{b_3}); \quad (2^{a_1} \cdot 5^{b_1}) \cdot (2^{a_3} \cdot 5^{b_3}) \cdot (2^{a_2} \cdot 5^{b_2});$$
$$(2^{a_2} \cdot 5^{b_2}) \cdot (2^{a_1} \cdot 5^{b_1}) \cdot (2^{a_3} \cdot 5^{b_3}); \quad (2^{a_2} \cdot 5^{b_2}) \cdot (2^{a_3} \cdot 5^{b_3}) \cdot (2^{a_1} \cdot 5^{b_1});$$
$$(2^{a_3} \cdot 5^{b_3}) \cdot (2^{a_1} \cdot 5^{b_1}) \cdot (2^{a_2} \cdot 5^{b_2}); \quad (2^{a_3} \cdot 5^{b_3}) \cdot (2^{a_2} \cdot 5^{b_2}) \cdot (2^{a_1} \cdot 5^{b_1}).$$

Hence the total number of different decompositions of 1,000,000 into a product of three factors is

$$1 + 15 + \frac{784 - 15 \cdot 3 - 1 \cdot 1}{6} = 1 + 15 + \frac{738}{6} = 1 + 15 + 123 = 139.$$

18. The factorization of 18,000 into a product of primes is $18,000 = 2^4 \cdot 3^2 \cdot 5$. Hence all divisors of 18,000 have the form $2^a \cdot 3^b \cdot 5^c$, where a, b, c are integers satisfying $0 \leq a \leq 4$, $0 \leq b \leq 2$, $0 \leq c \leq 3$. (Note that the divisor 1 is gotten by taking $a = b = c = 0$.) Thus there are 5 possibilities for a (namely 0, 1, 2, 3, or 4), 3 possibilities for b, and 4 possibilities for c. Since these can be combined in all possible ways, the number of divisors of 18,000 is $5 \cdot 3 \cdot 4 = 60$.

Let us now find the *sum* of all the divisors. We want to sum the numbers $2^a \cdot 3^b \cdot 5^c$, where a, b, c range through the values specified above. If the expression $(1 + 2 + 2^2 + 2^3 + 2^4)(1 + 3 + 3^2)(1 + 5 + 5^2 + 5^3)$ is expanded in the usual way, its terms will be precisely these numbers $2^a \cdot 3^b \cdot 5^c$ (note that $2^0 = 3^0 = 5^0 = 1$). So the desired sum is

$$(1 + 2 + 2^2 + 2^3 + 2^4)(1 + 3 + 3^2)(1 + 5 + 5^2 + 5^3) = 31 \cdot 13 \cdot 156$$

$$= 62,868.$$

We present now a variant of this solution. Let m and n be two relatively prime positive integers. In this case every divisor of mn can be uniquely expressed as the product of a divisor of m and a divisor of n. It can be seen from this that if $\tau(N)$ denotes the number of divisors of the positive integer N, then

$$\tau(mn) = \tau(m)\tau(n).$$

Likewise if $\sigma(n)$ is the sum of the divisors of N, we have

$$\sigma(mn) = \sigma(m)\sigma(n).$$

In number theory a function $f(N)$ of the positive integer N is called *multiplicative* if $f(mn) = f(m)f(n)$ whenever m and n are relatively prime. Another example of such a function is Euler's function $\varphi(N)$, the number of positive integers $\leq N$ and relatively prime to N (see the remark to problem 14).

Thus $\tau(N)$ and $\sigma(N)$ are multiplicative functions. Note, however, that if m and n are not relatively prime, then the equations $\tau(mn) = \tau(m)\tau(n)$ and $\sigma(mn) = \sigma(m)\sigma(n)$ need not hold; for example, $\tau(4) = 3$, but $\tau(2)\tau(2) = 2 \cdot 2 = 4$. If $f(N)$ is multiplicative, then $f(n_1 n_2 \cdots n_k) = f(n_1)f(n_2) \cdots f(n_k)$, provided that every pair of the numbers $n_1, \ldots, n_k$ are relatively prime. Since $18,000 = 2^4 \cdot 3^2 \cdot 5^3$, this implies that $\tau(18,000) = \tau(2^4)\tau(3^2)\tau(5^3)$.

Now if p is a prime, then $\tau(p^r) = r + 1$, since the divisors of p^r are $1, p, p^2, \ldots, p^r$. Hence $\tau(18,000) = 5 \cdot 3 \cdot 4 = 60$. Similarly we see that $\sigma(18,000) = \sigma(2^4)\sigma(3^2)\sigma(5^3)$. Since

$$\sigma(p^r) = 1 + p + p^2 + \cdots + p^r = \frac{p^{r+1} - 1}{p - 1},$$

we have $\sigma(18,000) = 31 \cdot 13 \cdot 156 = 62,868$.

Remark. The same reasoning leads immediately to the fact that if N has a factorization into prime factors of the form

$$N = p_1{}^{a_1}p_2{}^{a_2} \cdots p_k{}^{a_k},$$

then
$$\tau(N) = (a_1 + 1)(a_2 + 1) \cdots (a_k + 1),$$

and
$$\sigma(N) = \frac{p_1{}^{a_1+1} - 1}{p_1 - 1} \frac{p_2{}^{a_2+1} - 1}{p_2 - 1} \cdots \frac{p_k{}^{a_k+1} - 1}{p_k - 1}.$$

19. The factorization of 126,000 into primes is $126,000 = 2^4 \cdot 3^2 \cdot 5^3 \cdot 7$. Consequently A and B must have the form

$$A = 2^{a_1} \cdot 3^{b_1} \cdot 5^{c_1} \cdot 7^{d_1}, \quad B = 2^{a_2} \cdot 3^{b_2} \cdot 5^{c_2} \cdot 7^{d_2}.$$

Furthermore it is necessary that

$$\max(a_1,a_2) = 4, \qquad \max(b_1,b_2) = 2,$$
$$\max(c_1,c_2) = 3, \qquad \max(d_1,d_2) = 1,$$

where $\max(x,y)$ denotes the greater of the two numbers x and y.

There are 9 ordered pairs (a_1,a_2) with $\max(a_1,a_2) = 4$; indeed there are 5 such pairs with $a_1 = 4$ (since a_2 can then be 0, 1, 2, 3, or 4), and 5 pairs with $a_2 = 4$, but the pair $(4,4)$ has been counted twice, so we get a total of $5 + 5 - 1 = 9$. Similarly there are 5 pairs (b_1,b_2) with $\max(b_1,b_2) = 2$, there are 7 pairs (c_1,c_2) and 3 pairs (d_1,d_2). Combining these in all possible ways, we get a total of $9 \cdot 5 \cdot 7 \cdot 3 = 945$ pairs (A,B).

However, in this enumeration we have been counting pairs of numbers which differ only in the order of the two components as different. In fact every pair (A,B) with 126,000 as least common multiple *except the pair* $A = B = 126,000$ was counted twice. Hence the number of different pairs (A,B) is $(945 - 1)/2 + 1 = 473$.

Remark. It can be shown in the same way that if the factorization of a positive integer N into primes is

$$N = p_1{}^{m_1}p_2{}^{m_2} \cdots p_k{}^{m_k},$$

then the number of different pairs (A,B) whose least common multiple is N, is

$$\frac{(2m_1 + 1)(2m_2 + 1) \cdots (2m_k + 1) - 1}{2} + 1.$$

20. We can write

$$(1 + x^5 + x^7)^{20} = (1 + x^5 + x^7)(1 + x^5 + x^7) \cdots (1 + x^5 + x^7),$$

where there are twenty factors on the right. A typical term in the expansion of $(1 + x^5 + x^7)^{20}$ is obtained by selecting from each factor on the right either 1 or x^5 or x^7, and then multiplying the selected terms together. For example, if we took x^5 from the first factor, x^7 from the second factor, and 1 from all the other factors we would get the term $x^5 \cdot x^7 = x^{12}$. The power of x in such a product will be a sum of 5's and 7's. But the number 18 cannot be represented as a sum of 5's and 7's in any way; hence there is no term in x^{18} in the expansion (that is, the coefficient of x^{18} is zero).

The number 17 can be written in exactly one way as a sum of 5's and 7's: $17 = 5 + 5 + 7$. Consequently, the coefficient of x^{17} is equal to the number of terms obtained by selecting x^5 from two of the factors $(1 + x^5 + x^7)$, x^7 from one of them, and 1 from the remaining 17. The x^7 can be selected from any of the 20 factors $(1 + x^5 + x^7)$. Let us discuss for definiteness the case in which it is selected from the first factor. Then there are 19 remaining factors. From two of these we must select x^5; this can be done in $\binom{19}{2} = 171$ ways. The same holds for each of the 20 possible positions of the x^7, so we get a total of $20 \cdot 171 = 3420$ terms in x^{17}. Therefore the coefficient of x^{17} is 3420.

21. In changing the quarter we can use either 2 dimes or 1 dime or no dimes. If 2 dimes are used, then the change can be completed by either a nickel or by 5 pennies. Thus we get 2 possibilities in this case.

If one dime is used, the remaining 15¢ must be made up out of nickels and pennies. This can be done by using either 3, 2, 1, or no nickels (the remainder being made up of pennies in each case). Hence we have 4 possibilities involving one dime.

Finally, if no dimes are used, we can use either 5, 4, 3, 2, 1, or no nickels (the remainder of the 25¢ consisting of pennies in each case). Hence we have 6 possibilities involving no dimes.

The total number of ways of changing the quarter is therefore $2 + 4 + 6 = 12$.

22. The number of nickels used in putting together n cents cannot exceed $\left[\dfrac{n}{5}\right]$. On the other hand if q is any integer in the range $0 \leq q \leq \left[\dfrac{n}{5}\right]$, then we can put together q nickels and $n - 5q$ pennies to make up n cents. The solution is therefore $\left[\dfrac{n}{5}\right] + 1$.

Remark. The same reasoning shows that if the nickel were replaced by another coin having the value of k pennies, then the number of ways to put together n cents using pennies and the new coins would be $\left[\dfrac{n}{k}\right] + 1$.

23a. In making up n cents out of 1-, 2-, and 3-cent stamps, we can use either no 3-cent stamps at all, or one 3-cent stamp, or 2 of them, etc., up to a maximum of $q = [n/3]$ of them. In the first case the n cents would have to be made up entirely out of 1- and 2-cent stamps, which can be done in $[n/2] + 1$ ways. (See the remark to problem 22 with $k = 2$.) In the second case we must form $n - 3$ cents with 1- and 2-cent stamps, which can be done in $[(n - 3)/2] + 1$ ways; in the third case we must form $n - 6$ cents, which can be done in $[(n - 6)/2] + 1$ ways, etc. Let $n = 3q + r$, where $q = [n/3]$ is the quotient obtained by dividing 3 into n, and r is the remainder (thus $r = 0, 1,$ or 2). We then see that in the final case (where q 3-cent stamps are used), the remaining r cents must be made up out of 1- and 2-cent stamps. This can be done in $[r/2] + 1$ ways, so we get a total of

$$S = \left(\left[\frac{n}{2}\right] + 1\right) + \left(\left[\frac{n-3}{2}\right] + 1\right)$$
$$+ \left(\left[\frac{n-6}{2}\right] + 1\right) + \cdots + \left(\left[\frac{r}{2}\right] + 2\right)$$

as the solution to the problem.

We will now show that this sum is equal to the nearest integer to $(n + 3)^2/12$, that is, $S = N((n + 3)^2/12)$ using the notation introduced on page 6). First of all, we note that for any integer m,

$$\left[\frac{m}{2}\right] = \frac{m}{2} - \frac{1}{4} + \frac{(-1)^m}{4}.$$

Indeed, if m is even, then both the left side and right side are equal to $m/2$, and if m is odd, both sides equal $(m - 1)/2$.

This fact can be used to simplify the expression for S. We have

$$S = \left(\frac{n}{2} + \frac{3}{4} + \frac{(-1)^n}{4}\right) + \left(\frac{n-3}{2} + \frac{3}{4} + \frac{(-1)^{n-3}}{4}\right)$$
$$+ \left(\frac{n-6}{2} + \frac{3}{4} + \frac{(-1)^{n-6}}{4}\right) + \cdots + \left(\frac{r}{2} + \frac{3}{4} + \frac{(-1)^r}{4}\right).$$

Since there are $q + 1$ parentheses each containing a $\frac{3}{4}$, we have

$$S = (q + 1)\tfrac{3}{4} + \left(\frac{(-1)^n}{4} + \frac{(-1)^{n-3}}{4} + \frac{(-1)^{n-6}}{4} + \cdots + \frac{(-1)^r}{4}\right)$$
$$+ \left(\frac{n}{2} + \frac{n-3}{2} + \frac{n-6}{2} + \cdots + \frac{r}{2}\right).$$

The terms in the first parenthesis alternate in sign, since $(-1)^3 = -1$. Therefore this parenthesis is equal to $\frac{1}{4}$ if both n and r are even, $-\frac{1}{4}$ if both are odd, and 0 otherwise. We will denote its value by ε, and note for later purposes that $\varepsilon \geq 0$ when r is even.

The terms of the second parenthesis are in arithmetic progression. Recalling that the sum of such a progression is the average of the first and last terms multiplied by the number of terms (which in this case is $q + 1$), we see that

$$S = (q+1)\tfrac{3}{4} + \varepsilon + \frac{q+1}{2}\left(\frac{n}{2} + \frac{r}{2}\right)$$

$$= \frac{q+1}{4}(3 + n + r) + \varepsilon$$

$$= \frac{3q+3}{12}(n + 3 + r) + \varepsilon$$

$$= \frac{(n+3-r)(n+3+r)}{12} + \varepsilon$$

$$= \frac{(n+3)^2}{12} - \frac{r^2}{12} + \varepsilon.$$

Now $|-(r^2/12) + \varepsilon| < \tfrac{1}{2}$, because if $r = 0$ or 1, then

$$\left|-\frac{r^2}{12} + \varepsilon\right| \leq \frac{1}{12} + \frac{1}{4} < \frac{1}{2},$$

and if $r = 2$, we know from the above that $\varepsilon \geq 0$, so that

$$\left|-\frac{r^2}{12} + \varepsilon\right| \leq -\frac{1}{12} + \frac{1}{4} < \frac{1}{2}.$$

Thus S, which is an integer, differs from $(n+3)^2/12$ by a quantity whose absolute values is less than $\tfrac{1}{2}$. Therefore $S = N((n+3)^2/12)$.

23b. The solution to this problem is completely analogous to that of problem 23a. Let $n = 5q + r$ where $q = [n/5]$, and $r = 0, 1, 2, 3,$ or 4; thus q and r are the quotient and remainder when n is divided by 5. The number of 5-cent stamps used in making up n cents can vary from 0 to q; as in part a we find that if t 5-cent stamps are used, the number of solutions is $[(n - 5t)/2] + 1$. Hence the total number of solutions is

$$S = \left(\left[\frac{n}{2}\right] + 1\right) + \left(\left[\frac{n-5}{2}\right] + 1\right) + \left(\left[\frac{n-10}{2}\right] + 1\right) + \cdots$$
$$+ \left(\left[\frac{r}{2}\right] + 1\right).$$

We will now prove that $S = N((n + 4)^2/20)$. As in part a the identity $[m/2] = m/2 - \frac{1}{4} + (-1)^m/4$ shows that

$$S = (q + 1)\tfrac{3}{4} + \left(\frac{(-1)^n}{4} + \frac{(-1)^{n-5}}{4} + \cdots + \frac{(-1)^r}{4}\right)$$
$$+ \left(\frac{n}{2} + \frac{n - 5}{2} + \cdots + \frac{r}{2}\right).$$

Again the terms of the first parenthesis alternate in sign, so the sum of these terms is $\frac{1}{4}$ if n and r are even, $-\frac{1}{4}$ if n and r are odd, and 0 otherwise. Calling this sum ε we note that $\varepsilon \geqq 0$ if r is even. The second parenthesis can be evaluated by the formula for the sum of an arithmetic progression as in part a; we get

$$S = (q + 1)\tfrac{3}{4} + \varepsilon + \frac{q + 1}{2}\left(\frac{n}{2} + \frac{r}{2}\right)$$
$$= \frac{q + 1}{4}(3 + n + r) + \varepsilon$$
$$= \frac{5q + 5}{20}(3 + n + r) + \varepsilon$$
$$= \frac{(n - r + 5)(n + r + 3)}{20} + \varepsilon$$
$$= \frac{(n + 4)^2}{20} - \frac{(r - 1)^2}{20} + \varepsilon.$$

Now if $r = 0$, 1, 2, or 3 then $|-(r - 1)^2/20 + \varepsilon| \leq \frac{4}{20} + \frac{1}{4} < \frac{1}{2}$, while if $r = 4$, then $\varepsilon \geqq 0$, so that $|-(r - 1)^2/20 + \varepsilon| \leq |-\frac{9}{20} + \frac{1}{4}| < \frac{1}{2}$. Hence, since S is an integer, $S = N((n + 4)^2/20)$.

24. Since only an integral multiple of \$10 can be made up out of 10, 20, and 50-dollar bills, the 1, 2, and 5-dollar bills must also add up to a multiple of \$10.

Suppose, therefore, that the 1's, 2's, and 5's add up to $10m$ dollars, and the 10's, 20's, and 50's to $100 - 10m$ dollars. Here m can range from 0 to 10. We will determine the number of solutions for each fixed value of m, and then add these together to get the answer to the problem.

By problem 23b, the number of ways of breaking the integer $10m$ into a sum of 1's, 2's, and 5's is $N(10m + 4)^2/20)$. On the other hand the problem of breaking $100 - 10m = 10(10 - m)$ into 10's, 20's, and 50's is equivalent to that of writing $10 - m$ as a sum of 1's, 2's, and 5's (because everything has merely been multiplied by 10). Hence the number of solutions is $N((10 - m + 4)^2/20) = N((14 - m)^2/20)$. Each of these solutions can be combined with each of the representations of $10m$ as a sum of 1's, 2's, and 5's; therefore with this value of m there are

$N((10m + 4)^2/20)N((14 - m)^2/20)$ ways of changing the \$100. Summing over the values of m from 0 to 10 we get a total of

$$N\left(\frac{16}{20}\right)N\left(\frac{196}{20}\right) + N\left(\frac{196}{20}\right)N\left(\frac{169}{20}\right) + N\left(\frac{576}{20}\right)N\left(\frac{144}{20}\right)$$

$$+ N\left(\frac{1156}{20}\right)N\left(\frac{121}{20}\right) + N\left(\frac{1936}{20}\right)N\left(\frac{100}{20}\right)$$

$$+ N\left(\frac{2916}{20}\right)N\left(\frac{81}{20}\right) + N\left(\frac{4096}{20}\right)N\left(\frac{64}{20}\right)$$

$$+ N\left(\frac{5476}{20}\right)N\left(\frac{49}{20}\right) + N\left(\frac{7056}{20}\right)N\left(\frac{36}{20}\right)$$

$$+ N\left(\frac{8836}{20}\right)N\left(\frac{25}{20}\right) + N\left(\frac{10,816}{20}\right)N\left(\frac{16}{20}\right)$$

$$= 1 \cdot 10 + 10 \cdot 8 + 29 \cdot 7 + 58 \cdot 6 + 97 \cdot 5$$
$$+ 146 \cdot 4 + 205 \cdot 3 + 274 \cdot 2 + 353 \cdot 2$$
$$+ 442 \cdot 1 + 541 \cdot 1$$
$$= 10 + 80 + 203 + 348 + 485 + 584 + 615$$
$$+ 548 + 706 + 442 + 541$$
$$= 4562.$$

25. If n is represented as a sum of two positive integers
$$n = x + y,$$
then one of the terms must be less than or equal to $n/2$. This term can take the values $1, 2, 3, \ldots, [n/2]$; all these cases are different, since the second term will in these cases be at least $n/2$. Hence there are $[n/2]$ such representations.

26. For any integer k, let S_k be the number of solutions of $|x| + |y| = k$. Then the answer to the problem is $S = S_0 + S_1 + S_2 + \cdots + S_{99}$. Now $S_0 = 1$, since $|x| + |y| = 0$ is satisfied only by $x = y = 0$. We will show that for $k \geqq 1$, $S_k = 4k$. Indeed, x must have one of the $2k + 1$ values $-k, -k + 1, \ldots, k$. When $x = -k$ or $x = k$, we have only one value for y, namely $y = 0$. But for each of the other $2k - 1$ values of x, there are two values for y. Hence
$$S_k = 1 + 1 + 2(2k - 1) = 4k, \text{ and so}$$
$$S = 1 + 4(1 + 2 + \cdots + 99) = 1 + 4\frac{99 \cdot 100}{2}$$
$$= 19{,}801.$$

27. The problem amounts to that of determining the number of positive integral solutions of the equation $x + y + z = n$.

Observe first of all that the equation $y + z = k$ (where k is a positive integer) has $k - 1$ positive integral solutions. In fact, in this case y can take any of the values $1, 2, \ldots, k - 1$ (y cannot take the value k, since then z would not be positive); the corresponding value of z determined from the equation will also be a positive integer in each of these cases.

Let us now turn to our equation

$$x + y + z = n.$$

x can take any of the values $1, 2, 3, \ldots, n - 2$ (it could not take any larger value, since for $x > n - 2$, the quantity $x + y + z$ would be at least $(n - 1) + 1 + 1$, which exceeds n). y and z satisfy the equation

$$y + z = n - x;$$

hence for any fixed value of x there are $n - x - 1$ possible pairs of values for y and z. It follows from this that the total number of different solutions is

$$(n - 2) + (n - 3) + \cdots + n - (n - 2) - 1$$

$$= (n - 2) + (n - 3) + \cdots + 1$$

$$= \frac{(n - 1)(n - 2)}{2}.$$

28a. Suppose $n = x + y + z$ where x, y, and z are nonnegative integers. We will compute the number of such representations satisfying the inequalities $x \geq y \geq z$. This gives the answer to the problem because any other representation can be reduced to one of this type by reordering its terms (which is allowed by the conditions of the problem).

We now introduce the quantities a, b, c defined by $a = z$, $b = y - z$, $c = x - y$. These numbers are nonnegative integers, and solving for x, y, z in terms of them we get $z = a$, $y = z + b = a + b$, $x = y + c = a + b + c$. Therefore $n = x + y + z = a + 2b + 3c$. This equation can be interpreted as a representation of n as the sum of a 1's, b 2's and c 3's. So corresponding to each representation $n = x + y + z$ we have associated a partition of n into a sum of 1's, 2's, and 3's. For example, to the representation $10 = 5 + 3 + 2$ would correspond the partition $10 = 1 + 1 + 2 + 3 + 3$, since $a = 2$, $b = 1$, $c = 2$. Conversely if a, b, c are any nonnegative integers with $n = a + 2b + 3c$, then putting $x = a + b + c, y = b + c, z = c$ gives $n = x + y + z$ and $x \geq y \geq z \geq 0$. The correspondence can be visualized geometrically as follows.

Draw a row of x dots, beneath it a row of y dots, and beneath this a row of z dots. The case for $x = 5$, $y = 3$, $z = 2$, $n = 10$ is shown:

$$\begin{matrix} \cdot & \cdot & \cdot & \cdot & \cdot \\ \cdot & \cdot & \cdot & & \\ \cdot & \cdot & & & \end{matrix}$$

Now each column contains either 1, 2, or 3 dots, thus giving us a partition of n into 1's, 2's, and 3's. The inverse operation is also seen at once from this picture.

From what we have shown it follows that the total number of representations $n = x + y + z$, $x \geqq y \geqq z \geqq 0$, is equal to the number of ways of breaking n into a sum of 1's, 2's, and 3's. But this is the number we determined in problem 23a, where it was shown to be $N((n + 3)^2/12)$.

28b. If $n = x + y + z$, where x, y, and z are positive integers, then $n - 3 = (x - 1) + (y - 1) + (z - 1)$, where now $x - 1$, $y - 1$, and $z - 1$ are nonnegative. Conversely each representation of $n - 3$ as a sum of 3 nonnegative integers gives rise to a representation of n as a sum of 3 positive integers, obtained by increasing each term by 1. This fact reduces our problem to part a where, however, n must be replaced by $n - 3$. Therefore the answer is $N(n^2/12)$.

29. Since $z = n - x - y$, a solution is completely determined once x and y are specified. Let us see what conditions x and y must satisfy in order that x, y, and $z = n - x - y$ will satisfy the given inequalities.

1. The inequality $z \leqq x + y$ yields $n - x - y \leqq x + y$, which is equivalent to $x + y \geqq n/2$.
2. The inequality $y \leqq x + z$ yields $y \leqq x + n - x - y$, which is equivalent to $y \leqq n/2$.
3. The inequality $x \leqq y + z$ yields $x \leqq y + n - x - y$, or $x \leqq n/2$.
4. The inequalities $x > 0$, $y > 0$, $z > 0$ now become $x > 0$, $y > 0$, $x + y < n$.

Consequently the problem is to determine the number of pairs of integers (x, y) with $0 < x \leqq n/2$, $0 < y \leqq n/2$, $n/2 \leqq x + y < n$. These conditions can be interpreted geometrically as follows. We draw a rectangular coordinate system in the plane as in fig. 25, and let (x, y) be the coordinates of a point in this system. The three lines whose equations are $x = n/2$, $y = n/2$, and $x + y = n/2$ form the shaded triangle, and our inequalities amount to the requirement that (x, y) must be inside this triangle or on its boundary, but must not be one of the vertices M, N, P. So we must determine the number of points with integer coordinates in this region (indicated by black dots in the figures). It is now necessary to distinguish two cases, according as n is even or odd.

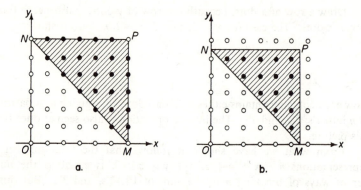

Fig. 25

Case 1: *n* even. Here $n/2$ is an integer, and we have the situation shown in fig. 25a (where $n = 12$). There are $n/2 + 1$ dots on the segment *OM*. Hence the total number of dots in the square *OMPN* is $(n/2 + 1)^2$. Of these points, $n/2 + 1$ are on the diagonal, and so the number of points to the right of the diagonal is $\frac{1}{2}\{(n/2 + 1)^2 - (n/2 + 1)\} = n(n + 2)/8$. Therefore the number of dots in the triangle *MNP* (including the vertices) is $n/2 + 1 + n(n + 2)/8$. Subtracting the 3 vertices we get

$$\frac{n}{2} - 2 + \frac{n(n + 2)}{8} = \frac{(n - 2)(n + 8)}{8}$$

as the answer.

Case 2: *n* odd. In this case $n/2$ is not an integer, so we have the situation in fig. 25b (where $n = 11$). Here the dots in the square *OMPN* form a smaller square *S*, and we want the number of dots to the right of the diagonal of *S*. The number of dots on *OM* is $1 + (n - 1)/2 = (n + 1)/2$, so *S* contains $[(n + 1)/2]^2$ dots. Of these, $(n + 1)/2$ are on the diagonal, and so

$$\frac{1}{2}\left\{\left(\frac{n + 1}{2}\right)^2 - \left(\frac{n + 1}{2}\right)\right\} = \frac{n^2 - 1}{8}$$

are to the right of the diagonal. (Note that the vertices *M*, *N*, *P* cause no trouble since they do not have integer coordinates.)

30. Denote the lengths of the sides of the triangle by x, y, and z. Then we must have $x + y + z = n$. Moreover, x, y, z must satisfy the inequalities

$$x < y + z, \quad y < x + z, \quad z < x + y,$$

since the length of a side of a triangle must be less than the sum of the lengths of the other two sides. Conversely, any quantities x, y, z which satisfy the above conditions will be the sides of a triangle of perimeter *n*. Thus the problem is similar to the preceding one, but differs from it in two respects. First, the required inequalities contain a $<$ sign where there was a $\leq$ sign before. Secondly, solutions differing only in the order of the terms must now be considered as the same, since they yield congruent triangles. For the moment we will ignore this second condition, and determine the number *N* of ordered triples x, y, z with $x + y + z = n$, $x < y + z$, $y < x + z$, $z < x + y$. As in problem 29, z is determined from x and y by the fact that $z = n - x - y$. The conditions on x and y now amount to $0 < x < n/2, 0 < y < n/2, n/2 < x + y < n$. This means that in fig. 25, the point (x,y) must be in the interior of the triangle MNP. Therefore, to find *N* we need merely subtract from the answer to problem 29 the number of dots on the boundary of MNP which were counted there. (Recall that the vertices were not counted in problem 29.) For odd *n* there are no dots on the boundary, so $N = (n^2 - 1)/8$. For even *n*, each side of the triangle contains, aside from the vertices, $n/2 - 1$ dots. Hence, in this case,

$$N = \frac{(n + 8)(n - 2)}{8} - 3\left(\frac{n}{2} - 1\right) = (n - 2)\frac{n + 8 - 12}{8}$$

$$= \frac{(n - 2)(n - 4)}{8} = \frac{n^2 - 6n + 8}{8}.$$

In this enumeration every scalene triangle Δ (that is, a triangle with no two sides equal) has been counted 6 times. For if the sides of Δ are p, q, r, then the solutions

1. $x = p, \quad y = q, \quad z = r$
2. $x = p, \quad y = r, \quad z = q$
3. $x = q, \quad y = p, \quad z = r$
4. $x = q, \quad y = r, \quad z = p$
5. $x = q, \quad y = r, \quad z = p$
6. $x = r, \quad y = q, \quad z = p$

all correspond to Δ.

Every proper isosceles triangle (two of whose sides have length p, while the third side has length q, where $q \neq p$) has been counted 3 times, for the solutions

1. $x = p, \quad y = p, \quad z = q$
2. $x = p, \quad y = q, \quad z = p$
3. $x = q, \quad y = p, \quad z = p$

all correspond to this triangle.

Finally, an equilateral triangle (if there is one) has been counted only once. Therefore $N = 6S + 3P + E$, where S is the number of scalene triangles, P is the number of proper isosceles triangles, and E is the number of equilateral triangles. Our problem is to compute $S + P + E = T$, the total number of triangles. From the formula $N = 6S + 3P + E$ we see that $S = (N - 3P - E)/6$; therefore

$$T = \frac{N - 3P - E}{6} + P + E = \frac{N + 3P + 5E}{6} = \frac{N + 3I + 2E}{6},$$

where $I = P + E$ is the total number of isosceles triangles (proper or equilateral). Since we have already found an expression for N, it remains only to compute E and I. Now $E = 1$ if n is a multiple of 3 and $E = 0$ otherwise (since an equilateral triangle of perimeter n has a side length of $n/3$). To find I we must determine the number of solutions of the equation $2x + y = n$ which satisfy the conditions $x > 0$, $y > 0$, $2x > y$. Putting $y = n - 2x$, these conditions become $x > 0$, $n - 2x > 0$, $2x > n - 2x$, or equivalently $n/4 < x < n/2$.

Since n and x are integers, these inequalities are equivalent to $n/4 < x \leq (n - 1)/2$. There are $[(n - 1)/2] - [n/4]$ such values of x, namely $x = [n/4] + 1, [n/4] + 2, \ldots, [(n - 1)/2]$. Hence

$$I = \left[\frac{n - 1}{2}\right] - \left[\frac{n}{4}\right].$$

We can now collect the above results to obtain a formula for T. It is convenient to write $n = 12q + r$, where $q = [n/12]$ is the quotient obtained by dividing n by 12, and the remainder r satisfies $0 \leq r \leq 11$. The formula for T depends on r as follows.

(1) If $r = 0$, then

$$N = \frac{n^2 - 6n + 8}{8}$$

$$I = \frac{n}{2} - 1 - \frac{n}{4} = \frac{n - 4}{4}$$

$$E = 1$$

$$T = \frac{N}{6} + \frac{I}{2} + \frac{E}{3} = \frac{n^2 - 6n + 8}{48} + \frac{n - 4}{8} + \frac{1}{3} = \frac{n^2}{48}.$$

(2) If $r = 1$, then

$$N = \frac{n^2 - 1}{8}$$

$$I = \frac{n - 1}{2} - \frac{n - 1}{4} = \frac{n - 1}{4}$$

$$E = 0$$

$$T = \frac{n^2 - 1}{48} + \frac{n - 1}{8} = \frac{n^2 + 6n - 7}{48}.$$

(3) If $r = 2$, then

$$N = \frac{n^2 - 6n + 8}{8}$$

$$I = \frac{n}{2} - 1 - \frac{n-2}{4} = \frac{n-2}{4}$$

$$E = 0$$

$$T = \frac{n^2 - 6n + 8}{48} + \frac{n-2}{8} = \frac{n^2 - 4}{48}.$$

(4) If $r = 3$, then

$$N = \frac{n^2 - 1}{8}$$

$$I = \frac{n-1}{2} - \frac{n-3}{4} = \frac{n+1}{4}$$

$$E = 1$$

$$T = \frac{n^2 - 1}{48} + \frac{n+1}{8} + \frac{1}{3} = \frac{n^2 + 6n + 21}{48}.$$

In the same way we obtain:

(5) If $r = 4$, then $T = \dfrac{n^2 - 16}{48}$

(6) If $r = 5$, then $T = \dfrac{n^2 + 6n - 7}{48}$

(7) If $r = 6$, then $T = \dfrac{n^2 + 12}{48}$

(8) If $r = 7$, then $T = \dfrac{n^2 + 6n + 5}{48}$

(9) If $r = 8$, then $T = \dfrac{n^2 - 16}{48}$

(10) If $r = 9$, then $T = \dfrac{n^2 + 6n + 9}{48}$

(11) If $r = 10$, then $T = \dfrac{n^2 - 4}{48}$

(12) If $r = 11$, then $T = \dfrac{n^2 + 6n + 5}{48}.$

Note that in all these cases the constant term of the polynomial in the numerator is less than half the denominator. We can therefore sum up the above results as follows:

If n is odd, $T = N((n^2 + 6n)/48)$; if n is even, $T = N(n^2/48)$.

31a. Consider a line segment of length n as shown in fig. 26. Any solution of the equation $x_1 + \cdots + x_m = n$ in positive integers corresponds to a decomposition of this segment into m pieces whose lengths are positive integers. The $m - 1$ end points of these pieces (other than P_o and P_n) must be chosen from among the $n - 1$ points $P_1, P_2, \ldots, P_{n-1}$ shown in the figure. This can be done in $\binom{n-1}{m-1}$ ways, and this is therefore the answer to the problem.

Fig. 26

31b. If $x_1 + \cdots + x_m = n$, where $x_1, \ldots, x_m$ are nonnegative integers, then $(x_1 + 1) + (x_2 + 1) + \cdots + (x_m + 1) = n + m$, where now $x_1 + 1$, $x_2 + 1, \ldots, x_m + 1$ are positive. Conversely, if $y_1 + \cdots + y_m = n + m$ with $y_1, \ldots, y_m$ positive, then $(y_1 - 1) + \cdots + (y_m - 1) = n$, where $y_1 - 1, \ldots, y_m - 1$ are nonnegative. So our problem is equivalent to part a with n replaced by $n + m$. Hence the answer is $\binom{n+m-1}{m-1}$.

32a. Consider any partition π: $n = n_1 + n_2 + \cdots + n_k$ with the parts arranged in decreasing order, so that $n_1 \geq n_2 \geq \cdots \geq n_k$. Draw a row of n_1 dots, beneath it a row of n_2 dots, etc., as shown below for the partition $10 = 5 + 2 + 2 + 1$.

This figure is called the *Ferrars graph* of the given partition π. If the graph is read by columns instead of rows we obtain a new partition π' of n, called the *conjugate* of π. For example, the figure above shows that the conjugate of $5 + 2 + 2 + 1$ is $4 + 3 + 1 + 1 + 1$.

If π is a partition of n into at most m parts, then its Ferrars graph will have at most m rows. Consequently the conjugate partition π' will have all its parts $\leq m$. The converse is also true, and so we have established a one-to-one correspondence between the two types of partitions.

Remark. For $m = 3$ the theorem just proved says that the number of partitions of n into at most 3 parts is equal to the number of partitions of n into 1's, 2's, and 3's. In problem 23a it was shown that this latter number is $N\left(\dfrac{(n + 3)^2}{12}\right)$.

32b. Let π: $n = x_1 + x_2 + \cdots + x_m$ be a partition of n into m distinct parts, arranged so that $x_1 > x_2 > \cdots > x_m$. Since $x_m \geqq 1$, we must have $x_{m-1} \geqq 2$, $x_{m-2} \geqq 3, \ldots$, and finally $x_1 \geqq m$. Therefore the Ferrars graph of π includes the triangular array of dots shown in fig. 27a (where $m = 4$). There are $1 + 2 + 3 + \cdots + m = m(m + 1)/2$ dots in this triangle.

Suppose now that this triangle is removed, and that the remaining dots are then shifted to the left as shown in fig. 27b. They then constitute a partition of $n - m(m + 1)/2$ into at most m parts. Conversely, given a partition of $n - m(m + 1)/2$ into at most m parts, we can adjoin these parts to the triangle, and thereby obtain a partition of n into m distinct parts.

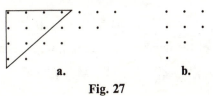

a.   b.

Fig. 27

33a. Any positive integer m can be written in the form $m = 2^k \cdot q$, where q is odd (if m is odd, then $k = 0$). Given a partition of n into distinct positive integers, we express each term in this form. Then we rearrange the terms, taking first those with $q = 1$ in increasing order, then those with $q = 3$ in increasing order, etc. Thus the partition becomes

$$n = 2^{a_1} + 2^{a_2} + \cdots + 2^{a_r} + 2^{b_1} \cdot 3 + 2^{b_2} \cdot 3 + \cdots + 2^{b_s} \cdot 3$$
$$+ 2^{c_1} \cdot 3 + 2^{c_2} \cdot 5 + \cdots + 2^{c_t} \cdot 5 + \cdots.$$

To this partition we can associate the partition of n into a sum of odd parts, consisting of $2^{a_1} + \cdots + 2^{a_r}$ 1's, $2^{b_1} + \cdots + 2^{b_s}$ 3's, $2^{c_1} + \cdots + 2^{c_t}$ 5's, etc. For example, to the partition

$$17 = 6 + 5 + 3 + 2 + 1 = 2 \cdot 3 + 5 + 3 + 2 \cdot 1 + 1$$
$$= 1 + 2 \cdot 1 + 3 + 2 \cdot 3 + 5$$

would be associated the partition

$$1 + 1 + 1 + 3 + 3 + 3 + 5.$$

Conversely, from a partition of n into odd parts we can determine uniquely a partition of n into distinct parts to which it is associated.

Suppose that in the given partition into odd parts there are A 1's, B 3's, C 5's, etc. Write the numbers $A, B, C, \ldots$ in the binary system, i.e. express them as sums of distinct powers of 2:

$$A = 2^{a_1} + 2^{a_2} + \cdots + 2^{a_r}$$
$$B = 2^{b_1} + 2^{b_2} + \cdots + 2^{b_s}$$
$$C = 2^{c_1} + 2^{c_2} + \cdots + 2^{c_t}, \text{ etc.}$$

Then the partition

$$n = 2^{a_1} + \cdots + 2^{a_r} + 2^{b_1} \cdot 3 + \cdots + 2^{b_s} \cdot 3 + 2^{c_1} \cdot 5 + \cdots$$
$$+ 2^{c_t} \cdot 5 + \cdots$$

is the only partition of n into distinct parts to which the given partition is associated under the correspondence described above.

33b. This is a generalization of part a and can be solved similarly. Let there be given a partition of a number n into parts which are not multiples of k; let s_1 be the number of times 1 occurs in this partition, $\ldots, s_{k-1}$ the number of times $k - 1$ occurs, s_{k+1} the number of times $k + 1$ occurs, etc.:

$$n = s_1 \cdot 1 + \cdots + s_{k-1}(k - 1) + s_{k+1}(k + 1) + \cdots$$

Now write each of the numbers $s_1, \ldots, s_{k-1}, s_{k+1}, \ldots$ in its k-ary form:

$$s_1 = q_0^{(1)} + q_1^{(1)} \cdot k + q_2^{(1)} \cdot k^2 + q_3^{(1)} \cdot k^3 + \ldots,$$
$$\cdot \quad \cdot \quad \cdot \quad \cdot \quad \cdot \quad \cdot \quad \cdot$$
$$s_{k-1} = q_0^{(k-1)} + q_1^{(k-1)}k + q_2^{(k-1)}k^2 + q_3^{(k-1)}k^3 + \cdots$$
$$s_{k+1} = q_0^{(k+1)}k + q_1^{(k+1)}k + q_2^{(k+1)}k^2 + q_3^{(k+1)}k^3 + \cdots$$
$$\cdot \quad \cdot \quad \cdot \quad \cdot \quad \cdot \quad \cdot \quad \cdot$$

where the "k-ary digits" $q_0^{(1)}, q_1^{(1)}, q_2^{(1)}, q_3^{(1)}, \ldots$, etc., can assume the values $0, 1, 2, \ldots, k - 1$. In this case we can associate to the given partition of n a new partition in which no part occurs more than $k - 1$ times:

$$n = \underbrace{1 + \cdots + 1}_{q_0^{(1)} \text{ terms}} + \underbrace{k + \cdots + k}_{q_1^{(1)} \text{ terms}} + \underbrace{k^2 + \cdots + k^2}_{q_2^{(1)} \text{ terms}} + \cdots$$

$$+ \underbrace{2 + \cdots + 2}_{q_0^{(2)} \text{ terms}} + \underbrace{2k + \cdots + 2k}_{q_1^{(2)} \text{ terms}} + \underbrace{2k^2 + \cdots + 2k^2}_{q_2^{(2)} \text{ terms}} + \cdots$$

$$\cdot \quad \cdot \quad \cdot \quad \cdot \quad \cdot \quad \cdot$$

$$+ \underbrace{(k - 1) + \cdots + (k - 1)}_{q_0^{(k-1)} \text{ terms}} + \underbrace{(k - 1)k + \cdots + (k - 1)k}_{q_1^{(k-1)} \text{ terms}}$$

$$+ \underbrace{(k + 1)k^2 + \cdots + (k + 1)k^2}_{q_2^{(k+1)} \text{ terms}} + \cdots.$$

Conversely, from any partition of n into parts none of which occurs more than $k - 1$ times, one can recover the original partition of n into parts not divisible by k. It follows from this that for any n there are as many partitions of the one kind as of the other.

Second solution. We present here another essentially different proof of the theorems of 33a and 33b; it suffices to consider part b, since part a is a special case of it.

Let A be the set of all partitions of n in which at least one part is divisible by k, and let B be the set of all partitions of n in which some part occurs at least k times. It suffices to show that $\#(A) = \#(B)$, and we will do this by using the principle of inclusion and exclusion to compute $\#(A)$ and $\#(B)$.

Let A_r be the set of all partitions of n in which the integer rk occurs as one of the parts. Then $A = A_1 \cup A_2 \cup A_3 \cup \cdots$, where there are actually only a finite number of non-empty terms on the right (since A_r is empty as soon as $rk > n$). By the principle of inclusion and exclusion,

$$
\begin{aligned}
\#(A) = \#(A_1) + \#(A_2) + \#(A_3) + \cdots \\
- \#(A_1 \cap A_2) - \#(A_1 \cap A_3) - \cdots \\
+ \#(A_1 \cap A_2 \cap A_3) + \cdots \\
- \cdots .
\end{aligned}
\tag{1}
$$

Similarly, let B_r be the set of all partitions of n in which the integer r occurs at least k times among the parts. Then $B = B_1 \cup B_2 \cup B_3 \cup \cdots$, and

$$
\begin{aligned}
\#(B) = \#(B_1) + \#(B_2) + \#(B_3) + \cdots \\
- \#(B_1 \cap B_2) - \#(B_1 \cap B_3) - \cdots \\
+ \#(B_1 \cap B_2 \cap B_3) + \cdots \\
- \cdots .
\end{aligned}
\tag{2}
$$

We next show that $\#(A_r) = \#(B_r)$. Given any partition $n = kr + x_1 + x_2 + \cdots + x_m$ of the set A_r we can associate to it the partition

$$
n = \overbrace{r + r + \cdots + r}^{k \text{ times}} + x_1 + x_2 + \cdots + x_m
$$

of the set B_r, and conversely. This sets up a one-to-one correspondence between the elements of A_r and B_r; hence $\#(A_r) = \#(B_r)$. By the same reasoning

$$
\#(A_r \cap A_s) = \#(B_r \cap B_s), \quad \#(A_r \cap A_s \cap A_t) = \#(B_r \cap B_s \cap B_t),
$$

etc. Therefore all terms on the right-hand side of (1) are equal to the corresponding terms of (2), from which it follows that $\#(A) = \#(B)$.

III. COMBINATORIAL PROBLEMS ON THE CHESSBOARD

34a. An $n \times n$ chessboard (see fig. 28, where the case $n = 8$ is illustrated) has n rows and n columns. For none of the rooks on the board to control the square on which another lies, it is necessary and sufficient that no two rooks lie in the same row or in the same column. Hence the total number of rooks cannot exceed n; on the other hand, it is possible for n rooks to be arranged on the board in such a way that none of them controls the square on which another one lies: for example, it would suffice to put them on the squares of one of the principal diagonals of the chessboard.

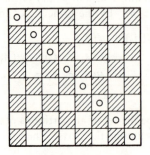

Fig. 28

We will now determine how many different arrangements of n rooks satisfy our conditions. Let us call the rook in the first column the first rook, that in the second column the second rook, . . . , and that in the n-th column the n-th rook. The first rook can lie in any of the n rows. For any choice of the row in which it lies, there remain $n - 1$ possible rows in which the second rook could lie (namely, all rows except that in which the first rook lies); once the locations of the first and second rooks are chosen, there remain $n - 2$ possible rows in which the third rook could lie, etc. Finally, there will be only one possibility left for the last rook once the locations of the first $n - 1$ have been specified. Pairing each of the n different possibilities for the first rook with each of the corresponding $n - 1$ possibilities for the second rook, we obtain a total of $n(n - 1)$ possible arrangements of the first two rooks; arguing similarly, there will be $n(n - 1)(n - 2)$ possible arrangements of the first three rooks, $n(n - 1)(n - 2)(n - 3)$ possible arrangements of the first four rooks, . . . , and finally $n(n - 1)(n - 2) \cdots 2 \cdot 1$ different arrangements of all n rooks.

Thus the total number of admissible arrangements of the rooks is

$$1 \cdot 2 \cdot 3 \cdots (n - 1)n = n!$$

In particular, for an ordinary chessboard (that is, for $n = 8$), we obtain:

$$8! = 1 \cdot 2 \cdot 3 \cdot 4 \cdot 5 \cdot 6 \cdot 7 \cdot 8 = 40,320$$

different arrangements.

34b. It is impossible for less than n rooks to control all squares of an $n \times n$ chessboard. In fact, if there were less than n rooks on the board, there would be a column on which there was no rook and a row on which there was no rook; the square common to this row and column would then not be controlled by any of the rooks. On the other hand, it is obviously possible to arrange n rooks on the board in such a way that they control every square of the board (see, for example, fig. 28).

If n rooks on an $n \times n$ chessboard control every square of the board, then there is either one rook in each column or one rook in each row. For otherwise there would be a row and a column, neither of which contained any rooks; and the square common to this row and column would not be controlled by any of the rooks. Conversely, if there is either one rook in every row or one rook in every column, then these rooks will control the entire board. The number of ways of arranging n rooks, one in each column, is n^n. (The first rook can be placed on any of the n squares of the first column; no matter which square it is put on, the second rook can be put on any of the n squares of the second column, etc.) The number of ways in which n rooks can be arranged, one in each row, is likewise n^n. It would seem at first glance that the number of arrangements of n rooks for which the rooks controlled all squares of the board would be equal to $n^n + n^n = 2n^n$. But in this enumeration we have counted twice each arrangement of the rooks for which there is one rook in each column and simultaneously one rook in each row. Since the total number of such arrangements is $n!$ (see solution to part a), the correct answer is $2n^n - n!$

In particular, for an ordinary chessboard ($n = 8$), we obtain

$$2 \cdot 8^8 - 8! = 33,514,312$$

different arrangements.

35a. Consider the diagonals which run from lower left to upper right on an ordinary 8×8 chessboard (for short, we will call these the positive diagonals). There are 15 such diagonals: the 8 diagonals which begin on the squares of the first column and the 7 diagonals which begin on the bottom squares of the other seven columns (fig. 29). If no bishop controls the square on which another lies, then there cannot be more than one bishop on each of these diagonals; consequently, the total number of bishops cannot be more than 15. But not all 15 diagonals can be occupied, since the first and last of them each consist of a single square, and a bishop on either of these two squares would control the other square. Consequently, at most 14 of the 15 diagonals can be occupied; therefore the number of bishops in such an arrangement cannot exceed 14.

On the other hand, 14 bishops can be arranged in the required way, as is shown, for example, in fig. 29. Hence this is the greatest number of bishops which can be arranged on an 8×8 chessboard in such a way that no bishop lies on a square controlled by another. In the more general case of an $n \times n$ board, the same reasoning shows that the maximum number of bishops is $2n - 2$.

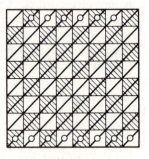

Fig. 29

35b. We will show that at least 8 bishops are needed to control all squares of an 8×8 board. To do this, we will treat the white and black squares separately, showing that there must be at least 4 bishops on each of the 2 colors. We will call a bishop black or white according as it is on a black or white square. If the black squares are rotated counterclockwise through an angle of 45° they assume the form of fig. 30a. It is convenient to transform this figure into fig. 30b, where the bishops now move horizontally and vertically, i.e. they have become rooks. The only reason for making this change is that the eye is apparently better able to visualize the moves of a rook than those of a bishop.

We now see that at least 4 rooks are needed to control the squares of fig. 30b since it contains a 4×4 square to which problem 34b can be applied. Thus we need at least 4 black bishops, and similarly we need at least 4 white bishops.

On the other hand, 8 bishops can in fact be arranged so as to control every square (see, for example, fig. 31a).

More generally, let $n = 2k$ be any even integer. Then we can prove that at least n bishops are needed to control the $n \times n$ chessboard. For if the black squares are rotated through 45° as before, and then redrawn so

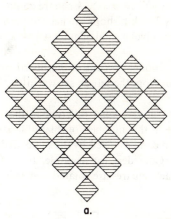

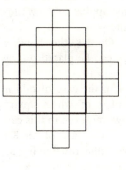

a.

b.

Fig. 30

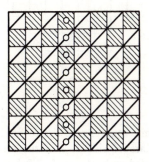

a.

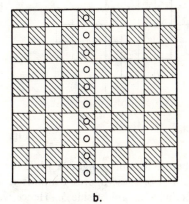

b.

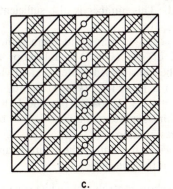

c.

Fig. 31

that the bishop becomes a rook, there will be a $k \times k$ square contained in the new figure. Therefore at least k black bishops are needed, and the same applies to the white bishops. This gives a total of at least $k + k = n$ bishops.

On the other hand, n bishops are enough to control the board, since they can be placed on the k-th column from the left as shown in fig. 31b for $n = 10$.

In the case of odd n the situation is somewhat different in that the number of white squares differs from the number of black squares. However, even in this case the problem can be solved by the same method as in the case of an 8×8 board. Consider, for example, the 9×9 board illustrated in fig. 31c. If the black squares are rotated and redrawn

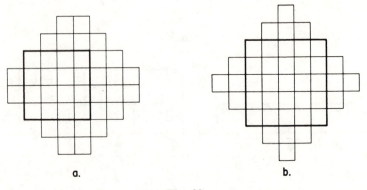

a.   b.

Fig. 32

as before we obtain fig. 32a while the white squares give rise to fig. 32b. The first of these figures contains a square of side 4, and the second contains a square of side 5. Hence we need at least $4 + 5 = 9$ bishops to control the board. This number is also sufficient, since 9 bishops can be arranged as in fig. 31c.

In general, if $n = 2k + 1$ is any odd number, then the same reasoning shows that a total of $k + (k + 1) = n$ bishops are needed. This is because for even k the transformed white squares (white being the color of the corners) contain a $(k + 1) \times (k + 1)$ board, and the transformed black squares contain a $k \times k$ board. For odd k the opposite holds, i.e., the transformed white squares contain a $k \times k$ board and the transformed black squares contain a $(k + 1) \times (k + 1)$ board. A total of n bishops is also sufficient to control the entire board, since for example they can be placed on the middle column.

36a. Since a white bishop controls only white squares and a black bishop controls only black squares, the problem of constructing a maximal arrangement of bishops none of which lies on a square controlled by another can be split into two independent parts; to construct a maximal arrangement of white bishops such that no bishop lies on a square controlled by another, and the corresponding problem for the black bishops. But in the case of even n, the union of all the black squares on the board and the union of all the white squares are congruent: one can be obtained from the other by rotating the board through 90°. Therefore the number of white bishops in a maximal arrangement is equal to the number of black bishops. Since these two numbers add up to $2n - 2$ (by problem 35a), each must be equal to $n - 1$. We obtain all admissible arrangements of $2n - 2$ bishops on the board by pairing each admissible arrangement of $n - 1$ white bishops with each such arrangement of $n - 1$ black bishops; but this is just the square of the number of arrangements of $n - 1$ white bishops.

36b. The solution of this part is completely analogous to that of part a.

37a. Let there be given an arrangement of $2n - 2$ bishops on an $n \times n$ chessboard such that no bishop controls another (see problem 35a). On each square of the board write the number of bishops controlling that square. A square occupied by a bishop is marked with a 1, since we are making the convention that each bishop controls itself. No square is marked with a 0, since if there were such a square we could place a new bishop on it without attacking any of the others, thus contradicting the maximality of the given arrangement. The corner squares are marked with 1's, since there is only one diagonal through such a square and there can be only one bishop on it. The non-corner squares are marked with either a 1 or a 2, since there are two diagonals through such a square, and there can be at most one bishop on each of these.

Of the four corners of the board, at least two are not occupied by bishops; for if more than two were occupied, we would have two bishops attacking each other. Hence there are at least $2n$ squares marked with a 1 (the $2n - 2$ squares occupied by bishops and the corners not occupied by bishops). Let S be the sum of all the numbers written on the chessboard. Since at least $2n$ of the terms added to form S are 1's and the other $n^2 - 2n$ terms are at most 2, we have

$$S \leqq 2n + 2(n^2 - 2n) = n(2n - 2).$$

Now suppose that in the given arrangement there are B bishops on the boundary and I bishops in the interior of the board (thus $B + I = 2n - 2$). A bishop on the boundary always controls exactly n squares (including the one it occupies); for example, if a bishop lies on a square on the top or bottom row, it will control exactly one square in each column. On the other hand, a bishop on an interior square controls at least

$n + 1$ squares. But when a bishop attacks a square it contributes one unit towards the number written on that square; therefore we have

$$S \geq nB + (n + 1)I$$
$$= n(B + I) + I$$
$$= n(2n - 2) + I.$$

Comparing this with the upper bound we previously found for S, we get

$$n(2n - 2) + I \leq n(2n - 2),$$

and so $I \leq 0$. Of course this implies that $I = 0$, so that all of the bishops are on the boundary.

37b. Consider an arbitrary outer square other than a corner square of the board (for example, the square in the bottom row marked by a circle in

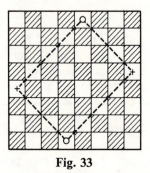

Fig. 33

fig. 33). Draw the two diagonals which pass through this square; these diagonals end at two other outer squares, which are marked by crosses in fig. 33. Now draw the other diagonals through the points marked by crosses; these diagonals will meet at an outer square of the board, the square which is symmetric through the center of the board to the original square (in fig. 33 this latter square is also marked with a circle).

Now consider any arrangement on the board of $2n - 2$ bishops in which no bishop lies on a square controlled by another. By virtue of the result of part a, all the bishops must be located on outer squares of the board. If one of the bishops is located in the bottom row but not in a corner (as marked by a circle in fig. 33), then the squares marked by crosses will be empty (since they are controlled by the bishop in the bottom row). Furthermore, there must be a bishop in the square marked with a circle in the top row, since otherwise there would be less than the maximum number of bishops, as a bishop could be placed on the latter square without controlling any of the others. Conversely, if one of the squares marked by a cross is occupied by a bishop, then the other is also occupied, and the two squares marked with circles are empty. Thus we have two possibilities for the arrangement of the bishops on the four marked outer squares: either there are bishops on the squares marked with circles and not on those marked with crosses, or vice versa.

Any of the $n - 2$ squares of the bottom row which are not corner squares can be taken as the one marked with a circle; thus we obtain $n - 2$ different rectangles of the sort constructed above (with vertices at the two crosses and the two circles). In each of these rectangles we must put a pair of bishops on opposite vertices. It is clear that for each rectangle, the choice of the pair of opposite vertices is independent of the choice for any other rectangle. By combining each of the two possible choices for the first rectangle with each of the two possible choices for the second, each of the two possible choices for the third, ..., and each of the two possible choices for the $(n - 2)$nd, we obtain a total of 2^{n-2} different possibilities. We now have only to consider the corner squares, since we have already taken care of all the bishops which do not lie on corner squares, of which there are a total of $2(n - 2)$. We are left with two bishops which we can put on corner squares in any way such that neither lies on a square controlled by the other; this means simply that the bishops are not on opposite ends of a diagonal, and there are four different arrangements which satisfy this condition (one bishop must lie on each of the principal diagonals, giving two choices for the one on the positive diagonal and two choices for the one on the negative diagonal). Combining these four ways of arranging bishops on the corner squares with the 2^{n-2} ways of arranging bishops on the other squares, we obtain a total of $4 \times 2^{n-2} = 2^n$ different arrangements of the $2n - 2$ bishops. Thus, the total number of admissible arrangements is 2^n.

In particular, for $n = 8$, we obtain a total of $2^8 = 256$ different arrangements.

Remark. The assertion of problem 36a follows from the answer to this problem; for even n, $2^{n/2}$ is an integer, and $2^n = (2^{n/2})^2$.

38a. It was shown in problem 35 that the minimum number of bishops needed to control the entire board is 8, of which 4 must be placed on black squares and 4 on white squares. If there are x ways in which the black squares can be controlled by 4 bishops and y ways in which the white squares can be controlled by 4 bishops, then the answer to the problem is xy (since each way of controlling the black squares can be combined with each way of controlling the white squares to produce a way of controlling the entire chess board). But $x = y$, since the union of the white squares is congruent to the union of the black squares. Therefore we need merely compute x, and the answer will then be x^2.

To calculate x it is convenient to make the same transformation as in problem 35, so that the black squares are as in fig. 34, and the bishops move horizontally and vertically, i.e., they have become rooks. The question is, in how many ways can 4 rooks be placed on this board so as to control it?

Suppose that one of the 4 middle rows did not contain a rook. Then all of its squares would have to be controlled vertically; but since there are at least 5 such squares, this is impossible. Hence there must be one rook on each of the 4 middle rows. Furthermore, there must be a rook on each of the 3 middle columns, for otherwise the 4 squares at the top and the 4 squares at the bottom would not be controlled. Conversely, if there is a rook on each of the 4 middle rows and each of the 3 middle columns, then the board is controlled.

We now have to distinguish two cases. In case 1, one rook is placed in the lightly shaded area A of fig. 34; the other three must then be in the

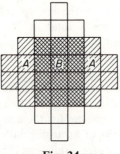

Fig. 34

heavily shaded rectangle B. Case 2 is where all four rooks are in B. In case 1 there are 12 ways to place a rook in area A; once it has been placed, the row on which it lies cannot contain another rook. Hence the other three rooks must be placed in B so as to occupy each column of B and each of the three remaining rows of B. This can be done in $3! = 6$ ways (see problem 34b). Therefore there are $12 \cdot 6 = 72$ ways in which case 1 can occur.

In case 2 one column of B must contain 2 rooks, and the other 2 columns contain one rook each. The column with 2 rooks can be selected in 3 ways, and once it is chosen, the rooks can be placed on it in $\binom{4}{2} = 6$ ways. For the other 2 rooks there remain 2 rows and 2 columns to be controlled, so they can be placed on the board in 2 ways. This gives a total of $3 \cdot 6 \cdot 2 = 36$ solutions under case 2. Hence $x = 72 + 36 = 108$, and $x^2 = 11{,}664$.

38b. As in part a we need merely compute the number x of ways in which the black squares can be controlled by 5 bishops; the solution is then x^2. Transforming the black squares as before, we obtain fig. 35, where the bishops are now replaced by rooks. If one of the 5 middle columns did not contain a rook, each of its squares would have to be controlled horizontally, which is impossible, since there are at least 6 such squares. Hence there is one rook on each of these columns. Also, each of the 4 middle rows must contain a rook in order to control the 6 leftmost squares and the 6 rightmost squares. Conversely, these conditions are sufficient to control the board. As before, there are 2 cases. In case 1 there is one rook in the lightly shaded area A and the other 4 are in the heavily shaded area B. Case 2 is where all 5 rooks are in B. In case 1 we can put a rook in A in 18 ways. This eliminates one column of B, leaving 4 rows and 4 columns to

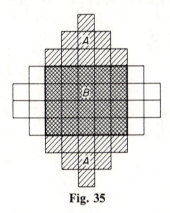

Fig. 35

be controlled. This can be done in $4! = 24$ ways by problem 34b. Hence we obtain $18 \cdot 24 = 432$ solutions in this case. In case 2, one row of B must contain 2 rooks and the other 3 rows contain 1 rook each. The row containing 2 rooks can be chosen in 4 ways; once it is chosen the 2 rooks can be placed on it in $\binom{5}{2} = 10$ ways. Then there are 3 rows and 3 columns left to be controlled by the other 3 rooks, so they can be distributed in $3! = 6$ ways. Thus we get $4 \cdot 10 \cdot 6 = 240$ solutions in this case. Hence $x = 432 + 240 = 672$, and $x^2 = 451,584$.

38c. We know from problem 35b that to control the 9×9 board of fig. 31c we must have 4 bishops on the black squares and 5 bishops on the white squares. Let x be the number of ways to control the black squares with 4 bishops, and y the number of ways to control the white squares with 5 bishops; then the answer to our problem is xy. As before, we transform the black squares to fig. 36a and the white squares to fig. 36b. To calculate x, notice that in fig. 36a each of the 4 middle columns must be occupied by a rook (for if such a column were empty its squares would have to be controlled horizontally, which is impossible since there are more than 4 such squares). Similarly each of the 4 middle rows must be occupied. Conversely, these conditions suffice to control the board. Hence, by problem 34b, $x = 4! = 24$.

The calculation of y is more difficult. In fig. 36b we see as before that each of the middle 3 rows must be occupied by a rook. Suppose that one of the two rows marked by arrows was not occupied. Then its squares would have to be controlled vertically, so that each of the middle 5 columns would have to be occupied. Thus we have proved that either the middle 5 rows are all occupied, or the middle 5 columns are all occupied (or both). Let a be the number of solutions in which the middle 5 rows are occupied, b the number of solutions in which the middle 5 columns are occupied, and

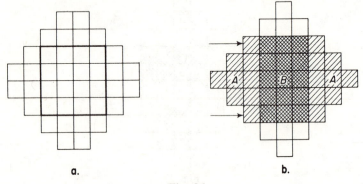

a.   b.

Fig. 36

c the number of solutions in which both of these conditions hold. Then $y = a + b - c$. By symmetry we see that $a = b$, and by problem 34b, $c = 5! = 120$. Hence $y = 2a - 120$, and we need only calculate a. When the middle 5 rows are occupied, the middle 3 columns must also be occupied in order to control the 4 squares at the top and the 4 squares at the bottom. Conversely, if these columns are occupied, the board is controlled. So the problem is: in how many ways can we place 5 rooks on the board of fig. 36b, with one rook on each of the 5 middle rows, so that each of the 3 middle columns contains at least one rook? This problem can be solved by using the principle of inclusion and exclusion. However, we will use instead the following method of solution, because although it is somewhat more cumbersome in the present case, it generalizes more readily to larger chessboards (see part d). We distinguish 4 different cases:

Case 1: All 5 rooks are in the heavily shaded rectangle B. One column of B contains 3 rooks, and the other 2 columns of B contain 1 rook each.

Case 2: All 5 rooks are in *B*. Two columns of *B* contain 2 rooks each, and the other column of *B* contains 1 rook.

Case 3: One rook is in the lightly shaded area *A*, and the other 4 are in *B*.

Case 4: Two rooks are in *A*, and the other 3 are in *B*.

In case 1 there are 3 ways to pick the column containing 3 rooks. Once it is chosen there are $\binom{5}{3} = 10$ ways to place the 3 rooks on it. Then there are $2! = 2$ ways to place the other 2 rooks on the other 2 columns (since only 2 rows remain for them). Hence there are $3 \cdot 10 \cdot 2 = 60$ solutions in this case.

In case 2, there are $\binom{3}{2} = 3$ ways to pick the 2 columns which contain 2 rooks each. Once these are picked there are $\binom{5}{2} = 10$ ways to put 2 rooks on the first of the chosen columns. Then there are $\binom{3}{2} = 3$ ways to put 2 rooks on the other chosen column (since only 3 of its rows are still usable). Finally, the position of the rook on the remaining column is completely determined. Hence we get $3 \cdot 10 \cdot 3 = 90$ solutions in this case.

In case 3, there are 18 ways to place a rook in the area *A*. Once it is placed it eliminates the row it is on from further use, thus leaving only 4 rows available for the rooks in *B*. Of these rooks, 2 must be on the same column, and 1 on each of the other 2 columns. There are 3 ways of picking the column with 2 rooks, and $\binom{4}{2} = 6$ ways of putting the rooks on it. Once these are placed there are $2! = 2$ ways of placing the last 2 rooks (since only 2 rows remain for them). Hence we get $18 \cdot 3 \cdot 6 \cdot 2 = 648$ solutions in this case.

In case 4, we must first pick 2 rooks in area *A*, not on the same row. The total number of ways to pick 2 squares of *A* (paying no attention to whether or not they are in the same row) is $\binom{18}{2} = 153$. From this we must subtract the number of pairs of squares of *A* which *are* in the same row in order to have left the number of pairs which are *not* in the same row. There is one way to pick 2 squares from the top row of *A*, $\binom{4}{2} = 6$ ways to pick 2 squares from the next row, $\binom{6}{2} = 15$ from the next, $\binom{4}{2} = 6$ from the next, and one from the bottom row of *A*. Hence the number of ways of putting 2 rooks in *A*, not on the same row, is $153 - 1 - 6 - 15 - 6 - 1 = 124$. Once these have been placed, there are only 3 rows left for the 3 rooks which must be placed in *B*. Hence these rooks can be positioned in

$3! = 6$ ways, by problem 34b. Thus we get $124 \cdot 6 = 744$ solutions in case 4. Combining the 4 cases we have

$$a = 60 + 90 + 648 + 744 = 1542$$

Therefore $y = 2 \cdot 1542 - 120 = 2964$, and $xy = 71{,}136$.

38d. The reasoning used in parts a, b, c can be generalized to an $n \times n$ chessboard. In the following discussion we let white be the color of the corners of the $n \times n$ board when n is odd, and define x and y as before. Thus x is the number of ways to control the black squares, and y the number of ways to control the white squares. There are 4 cases to consider.

Case 1: $n = 4k$. Here we need $2k$ black bishops and $2k$ white bishops. Reasoning as in part a, we find that $x = y = (2k)! \, (4k + 1)/2$.

Case 2: $n = 4k + 1$. Here there are $2k$ black bishops and $2k + 1$ white bishops. Reasoning as in part c we find for $k > 0$ that

$$x = (2k)!, \qquad y = 2\{(2k - 1)\binom{2k + 1}{3}(2k - 2)!$$

$$+ \binom{2k - 1}{2}\binom{2k + 1}{2}\binom{2k - 1}{2}(2k - 3)!$$

$$+ (k + 1)(2k + 2)(2k - 1)\binom{2k}{2}(2k - 2)!$$

$$+ \frac{2(k + 1)}{3}(3k^3 + 7k^2 + 5k)(2k - 1)!\} - (2k + 1)!$$

The 4 terms in the braces are obtained from the 4 cases analogous to those discussed in part c. The expression for y can be simplified to

$$y = \frac{16k^3 + 24k^2 + 11k + 1}{2}(2k)!$$

when $k = 0$, of course, $y = 1$.

Case 3: $n = 4k + 2$. Here there are $2k + 1$ black and $2k + 1$ white bishops. Reasoning as in part b we obtain

$$x = y = (4k^2 + 5k + 2)(2k)!$$

Case 4: $n = 4k + 3$. Here there are $2k + 2$ black and $2k + 1$ white bishops. By reasoning similar to that described above we find that

$$x = (16k^4 + 56k^3 + 67k^2 + 33k + 6)(2k)!$$
$$y = (2k + 1)!$$

39a. Divide the chessboard into 16 parts, each two squares wide and two squares high, as shown in fig. 37a. In an arrangement of kings such that none of them lies on a square controlled by another, none of these 16 parts can contain more than one king. It follows that it is impossible for

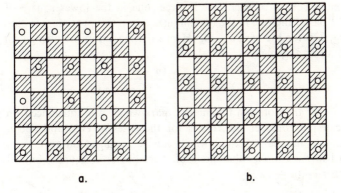

Fig. 37

more than 16 kings to be arranged on the board in such a way that none of them lies on a square controlled by another.

On the other hand, we can actually put 16 kings on the board in such a way that no one controls another; one way to do this is shown in fig. 37a. Consequently, the maximum number of kings in such an arrangement is 16.

39b. If n is even: $n = 2k$, then the problem can be solved exactly as in part a. Specifically, the board can be divided into $(n/2)^2 = k^2$ pieces 2 squares wide and 2 squares high. From the fact that none of these pieces can contain more than one king, it follows that the required maximum number of kings is at most k^2. But it is possible to arrange k^2 kings on a $2k \times 2k$ board in such a way that none of them is controlled by another; it suffices, for example, to put a king in the lower left-hand corner of each of the k^2 pieces into which we have divided the board. It follows from this that for n even, the required maximum number of kings is $k^2 = n^2/4$.

Now let n be odd: $n = 2k + 1$. We split the board into

$$\left(\frac{n+1}{2}\right)^2 = (k+1)^2$$

parts as indicated in fig. 37b (in such a decomposition we obtain k^2 two-by-two pieces, $2k$ pieces each consisting of two squares, and one piece which consists of a single square; that is, a total of $k^2 + 2k + 1 = (k+1)^2$ parts). It is clear that no more than one king can be put in any of these pieces. Consequently, the total number of kings cannot exceed $(k+1)^2$. But it is possible to arrange $(k+1)^2$ kings in such a fashion: put a king in the lower left-hand corner of each 2×2 piece, one in the left-hand square of each 1×2 piece, one in the lower square of each 2×1 piece, and one in the remaining square; this procedure is illustrated in fig. 37b for the case $k = 4$. Hence the maximum number of kings in such an arrangement is

$$(k + 1)^2 = \frac{(n + 1)^2}{4} .$$

Using the notation for integral part which was introduced on p. 6, we can put together the results for the cases of even and odd n: the maximum number of kings which can be arranged on an $n \times n$ chessboard in such a way that none of them lies on a square controlled by another is $[(n + 1)/2]^2$.

Remark. By considering fig. 37b, it is not hard to prove that for odd n there is *exactly one* arrangement of $(n + 1)^2/4$ kings on an $n \times n$ board for which none of the kings lies on a square controlled by another one. For even n (in particular, for $n = 8$) there are many different arrangements of $[(n + 1)/2]^2$ kings such that none of the kings attacks another. We leave it to the reader to compute the number of such arrangements.

40a. Divide the board into nine parts as indicated in fig. 38a. In each of these parts there is a square (marked by a circle in fig. 38a) which can be controlled only by a king which is on a square of the same part. Consequently, in order that every square be controlled by a king, it is necessary that there be at least one king in each of the nine parts. It follows from this that the required number of kings is at least nine. But it is possible to arrange nine kings on the board in such a way that they control the entire board: such an arrangement is illustrated, for example, by the circles in fig. 38a. Thus the minimum number of kings is nine.

40b. The problem can be solved in the same way as part a, but here we have to consider three cases separately: the case of n divisible by 3, the case where n leaves a remainder of 2 upon division by 3, and the case where n leaves a remainder of 1 upon division by 3.

If n is divisible by 3, that is, $n = 3k$, then the board can be divided into $k^2 = (n/3)^2$ pieces each 3 squares wide and 3 squares high (see fig. 38b); there must be a king in each of these pieces, since otherwise the middle square of one of the pieces would not be controlled by any king. Now $k^2 = n^2/9$ kings can always be arranged on a board of $n^2 = (3k)^2$ squares in such a way as to control all squares of the board (to do this, it suffices to put a king on the middle square of each of the k^2 parts into which the board has been divided; see fig. 38b). Hence the minimum number of kings is $k^2 = n^2/9$.

If n leaves a remainder of 2 upon division by 3, that is, $n = 3k + 2$, then the board can be divided into $(k + 1)^2 = [(n + 1)/3]^2$ parts in

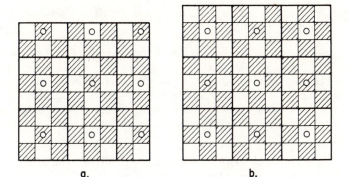

a. b.

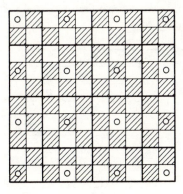

c.

Fig. 38

exactly the same way as in the case $n = 8$ (see fig. 38a). From a consideration of this decomposition, it follows that for $n = 3k + 2$ the minimum number of kings is

$$(k + 1)^2 = \frac{(n + 1)^2}{9}.$$

Finally, if n leaves a remainder of 1 upon division by 3, that is, $n = 3k + 1$, then the board can be divided into

$$(k + 1)^2 = \left(\frac{n + 2}{3}\right)^2$$

pieces in the way indicated in fig. 38c (where $n = 10$, $k = 3$). From a consideration of this figure it is seen that the minimum number of kings is

$$(k + 1)^2 = \frac{(n + 2)^2}{9}.$$

Using the symbol for integral part, the results obtained can be combined as follows: the minimum number of kings which can be arranged on an $n \times n$ chessboard in such a way as to control all squares of the board is $[(n + 2)/3]^2$.

Remark. It is not hard to see from a consideration of fig. 38b that for n divisible by 3 there is *exactly one* way in which $(n/3)^2$ kings can be arranged on a board of n^2 squares so as to control the entire board. For values of n which leave a remainder of 1 or 2 upon division by 3, $[(n + 2)/3]^2$ kings can be arranged on the board in such a way as to control all squares of the board in many different ways; we leave it to the reader to compute the number of such arrangements.

41a. There cannot be more than one queen in any column of the chessboard; hence it is impossible to arrange more than eight queens on an 8×8 chessboard in such a way that none of them lies on a square controlled by another.

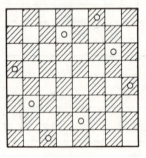

Fig. 39

On the other hand, we can actually put 8 queens on the board so as to satisfy this condition; one such arrangement is shown in fig. 39.

It can be shown that on an ordinary chessboard there are 92 different arrangements of eight queens which satisfy the condition imposed. (See, for example, M. Kraitchik, *Mathematical Recreations*, New York, 1942, p. 251.)

41b. There cannot be more than one queen in any column of the chessboard (since otherwise two queens would each control the square occupied by the other); hence it is impossible to arrange more than n queens on an $n \times n$ chessboard so as to satisfy the hypothesis of the problem.

If a single queen is put on a 2×2 chessboard, then it will control all squares of the board and thus no second queen can be put on the board (fig. 40a). On a 3×3 chessboard, one can arrange two queens so as to satisfy the hypothesis (fig. 40b), but it is impossible to do so with three queens. On a 4×4 or 5×5 chessboard it is possible to arrange four or five queens respectively, none of which lies on a square controlled by another (fig. 40c and d).

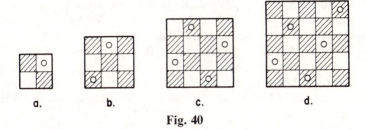

Fig. 40

We will now show that for $n \geqq 4$ it is possible to arrange n queens on an $n \times n$ chessboard so that none lies on a square controlled by another. Consider first the case of *even* $n = 2k$. In adjacent columns one cannot put two queens either in the same row or in adjacent rows (otherwise these queens would control each other horizontally or diagonally). We will therefore try putting each queen in a row two away from that in which we put the preceding one. Let us start by putting a queen on the second (that is, next to bottom) square of the first column, then on the fourth square of the second column, on the sixth square of the third column, etc., until we hit the top row of the board; then start again on the bottom square of the next column, then the third square of the next column, etc. (fig. 41). Since no two queens are in the same row or column, it remains only to prove that no two queens are on the same diagonal.

Let us treat separately the cases of positive and negative diagonals. The first $n/2 = k$ queens are arranged in such a way that the positive diagonal on which any of them lies is the one immediately above that on

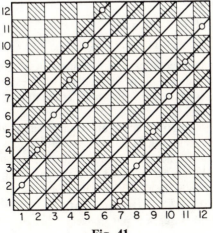

Fig. 41

which the previous one lies; similarly with the remaining k queens. Thus, the only way in which two of them could lie on the same positive diagonal would be for one of the first k queens to lie on the same positive diagonal as one of the second k queens. But this is impossible since the first k queens lie above the diagonal which joins the lower left-hand corner of the board to the upper right-hand corner, and the other k queens lie below this diagonal. Hence no two of the queens lie on the same positive diagonal.

If two squares of the board lie on the same negative diagonal, then the sum of the row number and the column number is the same for both of them. Conversely, if the sum of the row number and column number is the same for two squares, then they lie on the same negative diagonal. The row number of each of the squares on which the first k queens lie is twice the column number. The remaining k queens lie in the $(k + 1)$st through $2k$-th columns; the column number of the square in which one of these queens lies is thus of the form $k + s$, where s is a positive integer at most equal to k; it is not hard to see that the corresponding row number is $2s - 1$. For $r = 1, 2, \ldots, k$, the sum of the row and the column numbers of the square containing the r-th queen is $2r + r = 3r$; consequently, for each of the first k queens this sum has a different value, which means that no two of them lie on the same negative diagonal. Similarly, the sum of the row and column numbers for the $(k + s)$th queen $(s = 1, 2, \ldots, k)$ is $(2s - 1) + (k + s) = 3s + k - 1$, which takes a different value for each value of s; consequently, no two of the last k queens lie on the same negative diagonal. The only remaining possibility is that of one of the first k queens (say, the r-th) lies on the same negative diagonal as one of the last k queens (say, the $(k + s)$th). This will happen if and only if

$$3r = 3s + k - 1, \quad \text{that is,} \quad 3(r - s) + 1 = k = \tfrac{1}{2}n, \text{ or}$$
$$6(r - s) + 2 = n.$$

This is possible only when n leaves a remainder of 2 upon division by 6. Thus, for even n of the form $6m$ or $6m + 4$, fig. 41 gives an arrangement of n queens on the chessboard for which none of the queens lies on a square controlled by another.

For $n = 6m + 2$, fig. 41 leads to an arrangement in which two queens control each other. But even in this case we can find an arrangement of n queens, none of which lies on a square controlled by another, although this arrangement is more complicated than the preceding one. One such arrangement is shown in fig. 42 for the case of $n = 14$ (compare also with fig. 39). Here, in the $n/2 - 3$ columns starting with the 2nd and ending with the $(n/2 - 2)$nd, a queen is put in every other row starting with the 3rd (that is, the queen in the 2nd column lies in the 3rd row, that in the 3rd

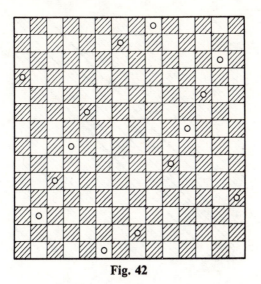

Fig. 42

column lies in the 5th row, that in the 4th column lies in the 7th row, etc.).
In the $n/2 - 3$ columns starting with the $(n/2 + 3)$rd and ending with the
$(n - 1)$st, queens are put in every other row, starting with the 6th (that is,
these queens are put respectively into the 6th, 8th, ..., and $(n - 2)$nd
rows). This leaves us with columns 1, $n/2 - 1$, $n/2$, $n/2 + 1$, $n/2 + 2$, and
n and rows 1, 2, 4, $n - 3$, $n - 1$, and n unoccupied. In the 1st, $(n/2 - 1)$st,
$(n/2)$nd, $(n/2 + 1)$st, $(n/2 + 2)$nd, and nth columns, the queens are
placed respectively in rows $n - 3$, 1, $n - 1$, 2, n, and 4. It is clear that no
two queens are in the same row or column; we thus have only to verify
that no more than one queen lies on any diagonal.

Let us label the positive diagonals by assigning the numbers 1 to
$2n - 1$ to the squares of the bottom row and the leftmost column as
indicated in fig. 43a and giving each positive diagonal the number of the
numbered square which belongs to it. We label the negative diagonals in a
way similar to this by assigning numbers to the squares of the bottom row
and the rightmost column as indicated in fig. 43b. If by the first queen we
mean that lying in the first column, by the second queen that lying in the
second column, etc. then the 1st, 2nd, 3rd, ..., and n-th queen will lie
respectively on the positive diagonals whose numbers are

$$2n - 4, n + 1, n + 2, n + 3, \ldots, 3n/2 - 3, n/2 + 2, 3n/2 - 1,$$

$$n/2 + 1, 3n/2 - 2, n/2 + 3, n/2 + 4, n/2 + 5, \ldots, n - 1, 4;$$

no two of these numbers will be equal provided that

$$4 < n/2 + 1, \qquad 2n - 4 > 3n/2 - 1,$$

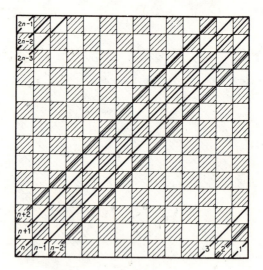

Fig. 43a

that is, if $n > 6$. Similarly, the queens lie respectively on the negative diagonals whose numbers are

$$n - 3, 4, 7, 10, 13, \ldots, 3n/2 - 8, n/2 - 1, 3n/2 - 2,$$
$$n/2 + 2, 3n/2 + 1, n/2 + 8, n/2 + 11, n/2 + 14,$$
$$n/2 + 17, \ldots, 2n - 4, n + 3,$$

where the dots denote terms of an arithmetic progression with difference 3. The numbers $4, 7, 10, 13, \ldots, 3n/2 - 8, 3n/2 - 2, 3n/2 + 1$ all give remainders of 1 on division by 3; $n/2 - 1, n/2 + 2, n/2 + 8, n/2 + 11, n/2 + 17, \ldots, 2n - 4$ are all divisible by 3 (recall that we are dealing with an n of the form $6m + 2$); $n - 3$ and $n + 3$ give remainders of 2 on division by 3. It is immediately clear from this that none of the numbers occurs more than once.

It now remains only to show that on an $n \times n$ board, where the number n is *odd* and $\geqq 5$, it is possible to arrange n queens in such a way that none of them lies on a square which another controls. But this becomes clear if one notes that in all the above arrangements constructed for even n, there are no queens on the diagonal joining the lower left-hand corner to the upper right-hand corner. Consequently, we can arrange n queens on an $n \times n$ board (n odd) in the following way: on the leftmost $n - 1$ columns and bottom $n - 1$ rows, $n - 1$ queens are arranged in such a way that none of them controls another according to the above scheme (this is possible since $n - 1$ is even), and the remaining queen is placed in the upper right-hand corner of the board. These n queens will

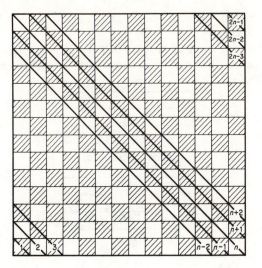

Fig. 43b

satisfy the required condition (see, for example, fig. 44 ,where an arrangement of 15 queens on a 15×15 board squares is illustrated).

To determine the *number* of different arrangements of n queens on an $n \times n$ board in which none of the queens lies on a square controlled by

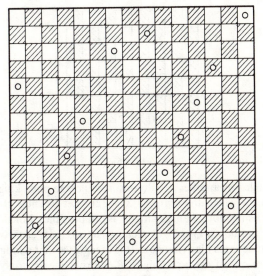

Fig. 44

another is extremely difficult, and so far no one has succeeded in doing so for the general case.

Another as yet unsolved problem is that of determining the minimum number of queens which can be arranged on an $n \times n$ chessboard so that

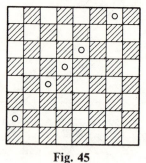

Fig. 45

they will control all squares of the board. For an ordinary 8×8 board, this number is 5 (see, for example, fig. 45); the number of different arrangements of 5 queens on an 8×8 board such that the queens control all squares of the board is 4860.

42a. Since a knight on a white square controls only black squares, it is obvious that 32 knights can be arranged in such a way that none lies on a

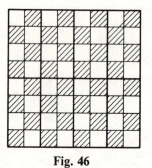

Fig. 46

square controlled by another; to do this, it suffices to put a knight on each white square (of which there are $64/2 = 32$). Let us show that such an arrangement using more than 32 knights is not possible. For this purpose, let us divide the board into eight rectangular sections, each two squares wide and four squares high (fig. 46). It is easy to see that a knight situated on a square of one of these rectangles R controls one and only one other square of R. Thus the squares of R can be divided into 4 pairs, and only one square of each pair can be occupied by a knight. It follows from this that no more than four knights can be arranged in one of these rectangles in such a way that none of them lies on a square controlled by another. Therefore, the total number of knights which can be arranged in such a way on the chessboard is at most $4 \times 8 = 32$.

42b. We must determine how many arrangements of 32 knights on a chessboard are such that none of them lies on a square controlled by another. Two such arrangements present themselves immediately: we can put the 32 knights on all the white squares of the board, or on all the black squares of the board. Let us prove that there are no other arrangements.

Divide the board once more into eight rectangular sections as indicated in fig. 46. On each section we must arrange exactly four knights (since we have 32 knights to dispose of and by the argument of part a, no more than four can be in any one section). Consider now how four knights can be arranged on the lower left-hand rectangle (we will call this the first rectangle).

Let us first try putting knights in each of the bottom two squares of this rectangle (these squares are marked by circles in fig. 47a). In this case we must leave empty the squares of the first rectangle which are marked by crosses: the two squares in the third row are controlled by the two knights, and the square in the second row marked with a cross must be left free since otherwise the three knights would control five squares of the second rectangle (that is, the one to the right of the first rectangle), and consequently it would be impossible to arrange four knights in that rectangle without one of them lying on a square controlled by another knight. Since the 2 squares marked with asterisks in fig. 47a cannot both be occupied, we must have a knight in the upper left-hand corner of the first rectangle. This leaves only two possible arrangements: those indicated by the circles in fig. 47b and 47c. If we arrange the knights on the squares of the first rectangle marked by circles in fig. 47b, then the squares of the second rectangle marked with circles will have to be the ones with knights on them (since the other four squares of the second rectangle are controlled by the knights in the first rectangle); but then only two knights could be put in the third rectangle (namely, on the squares marked with circles), since the other six squares are controlled by the knights in the second rectangle. Consequently, this possibility must be discarded. Finally, if we arrange the knights as in fig. 47c, then the knights in the second rectangle can be placed only in the first and fourth rows; then in the upper left rectangle, the knights can be placed only in the top two rows (since the other four squares of this rectangle are controlled by the four knights in the fourth row). But then the knights on the board control all squares of the second rectangle of the upper row, thus making it impossible to put any knights in that rectangle. We therefore see that the bottom two squares of the first rectangle cannot simultaneously be occupied.

If there were no knights in the bottom two squares of the first rectangle, then there would have to be knights on both squares of the third row. But these two knights would control five squares of the second rectangle; consequently, this arrangement is also inadmissible.

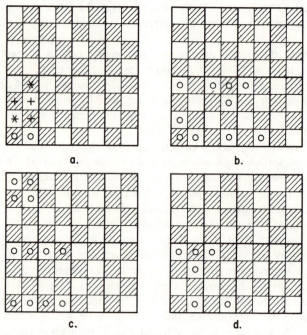

Fig. 47

Thus we are left with the case where there is exactly one knight in the bottom row of the first rectangle. There must likewise be exactly one knight in the top (fourth) row of this rectangle. For if there were no knights in the top row, there would have to be two knights in the second row, and these two knights would control five squares of the second rectangle. If there were two knights in the top row of the first rectangle, then either it would be impossible to arrange four knights in the second rectangle (fig. 47d) or it would be possible to arrange four knights in the second rectangle but not in the third (fig. 47e).

It is easy to see that the knight in the bottom row and the knight in the top row of the first rectangle must lie in different columns: otherwise there would again be no way to arrange four knights in each of the second and third rectangles (fig. 47f and g). But if these two knights lie in different columns (that is, if they lie on squares of the same color; see fig. 47h), then the two remaining knights can only be put on the other two squares of the same color; they are the only squares not controlled by the first two knights. Further, the only squares of the second rectangle which

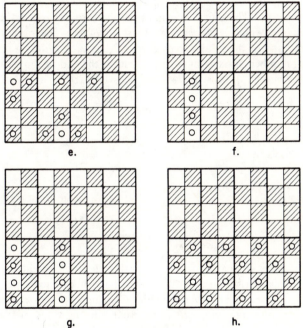

e.   f.

g.   h.

Fig. 47

are not controlled by the knights in the first rectangle are those of that same color; similar arguments for the third and fourth rectangles show that the knights in the lower half of the board must all lie on squares of the same color. Exactly the same argument is applicable to the upper half of the board, and thus all knights in the upper half of the board must lie on squares of the same color. Furthermore, if the knights in the bottom half of the board lay on squares of one color and those in the top half lay on squares of the other color, then some knights would lie on squares controlled by others. Therefore, there remain only two possibilities: the knights are either all put on white squares or all on black squares.

IV. GEOMETRIC PROBLEMS ON COMBINATORIAL ANALYSIS

43a. First draw the lines joining C to the n points on the side AB; these lines divide ABC into $n + 1$ smaller triangles. Now draw a line s from B to one of the points on AC (fig. 48). This line consists of $n + 1$ segments $s_1, \ldots, s_{n+1}$, and each of these segments cuts one of the small triangles into two pieces. Therefore, drawing s has increased by $n + 1$ the number of parts into which ABC was previously divided. The same argument applies to any of the n lines from B to AC; as each is drawn it increases the total number of parts by $n + 1$. Thus at the end of the process ABC is divided into $(n + 1) + n(n + 1) = (n + 1)^2$ parts.[1]

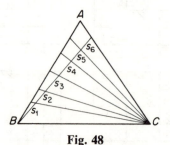

Fig. 48

43b. The lines emanating from the vertices B and C divide the triangle into $(n + 1)^2$ parts (see part a). Each of the lines emanating from the vertex A intersects all the $2n$ lines which emanate from the vertices B and C (the points of intersection are all different since no three of the lines are concurrent). Thus, each of the lines emanating from the vertex A is divided into $2n + 1$ pieces by the other lines and consequently increases the total number of pieces by $2n + 1$. It follows from this that the total number of pieces is

$$(n + 1)^2 + n(2n + 1) = 3n^2 + 3n + 1.$$

44a. It is clear that n lines will divide the plane into a maximum number of pieces if any two of these lines intersect (that is, if no two of them are parallel) and no three of them are concurrent. Consequently, we have

---

[1] This result can also be derived from the fact that the n lines emanating from the vertex B divide each of the $n + 1$ parts of the triangle already obtained into $n + 1$ smaller pieces. But the argument presented in the text is more general, and is applicable in many subsequent problems.

only to determine into how many pieces n mutually non-parallel lines, no three of which are concurrent, divide the plane.[2]

Suppose that k of the lines have already been drawn in the plane; let us draw the $(k + 1)$st line and see by how much it increases the number of pieces into which the plane is divided. The $(k + 1)$st line meets each of the k lines which have already been drawn; the k points of intersection divide it into $k + 1$ parts. Consequently, the $(k + 1)$st line cuts exactly $k + 1$ of the parts into which the plane has already been divided. Since it splits each of these parts into two pieces, drawing the $(k + 1)$st line increases the number of pieces by $k + 1$. But if only one line is drawn, it will divide the plane into two pieces. It follows from this

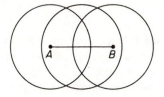

Fig. 49

that after n lines have been drawn the plane will have been divided into

$$2 + 2 + 3 + 4 + \cdots + n$$

parts (drawing the second line increases the number of parts by 2, drawing the third line increases it by 3 more, drawing the fourth increases it by 4 more, etc.). Consequently, the greatest number of parts into which n straight lines can divide the plane is

$$2 + 2 + 3 + \cdots + n = (1 + 2 + 3 + \cdots + n) + 1$$

$$= \frac{n(n + 1)}{2} + 1 = \frac{n^2 + n + 2}{2}.$$

44b. n circles will divide the plane into a maximum number of pieces if every two of them intersect (that is, if no two of them are tangent and none of them lies entirely within or outside of another) and no three of them are concurrent.[3]

---

[2] One might think that when these conditions are satisfied the number of pieces could still depend on the arrangement of the lines. However, it will follow from our proof that this number is unambiguously determined by the value of n and is consequently independent of the arrangement of the lines.

[3] Such sets of circles always exist; in fact it is possible to draw infinitely many circles in the plane in such a way that any two of them intersect in two points, but no three of them are concurrent. For example, construct two intersecting circles of the same radius r with centers A and B (see fig. 49). Then draw all circles of radius r whose centers are on the line segment AB. This family clearly has the desired properties.

By reasoning as in part a, we can show that the $(k + 1)$st circle increases by $2k$ the number of parts into which the plane is divided. For the $(k + 1)$st circle intersects each of the first k circles in two points; these $2k$ points divide the $(k + 1)$st circle into $2k$ arcs. Each of these arcs divides in two one of the regions formed by the first k circles. Since one circle divides the plane into two parts, the total number of parts after drawing the n-th circle is

$$2 + 2 + 4 + 6 + 8 + \cdots + 2(n - 1)$$
$$= 2 + 2(1 + 2 + 3 + \cdots + (n - 1))$$
$$= 2 + 2\,\frac{n(n - 1)}{2} = n^2 - n + 2.$$

45a. n planes will divide 3-dimensional space into a maximum number of parts if any three of them have exactly one point in common, and no four of them have a point in common.

Suppose that k planes have already been drawn; let us see by how much drawing the $(k + 1)$st plane increases the number of pieces into which space is divided. This plane meets each of the first k planes in a line; furthermore, any two of these lines of intersection have exactly one point in common (since any three of the planes have exactly one point in common), and no three of these lines are concurrent (since if three of them passed through the same point, at least four of the planes would pass through that point, which is excluded by the hypothesis). Consequently, these k lines divide the $(k + 1)$st plane into $(k^2 + k + 2)/2$ parts (the result of problem 44a), each of which is the surface on which the $(k + 1)$st plane meets one of the pieces formed by the first k planes. Thus the $(k + 1)$st plane cuts $(k^2 + k + 2)/2$ pieces, splitting each in two; consequently, drawing the $(k + 1)$st plane increases the number of pieces by $(k^2 + k + 2)/2$. Since a single plane divides space into two parts, it follows from this that the n planes will split space into

$$2 + \frac{1^2 + 1 + 2}{2} + \frac{2^2 + 2 + 2}{2} + \frac{3^2 + 3 + 2}{2} + \cdots$$
$$+ \frac{(n - 1)^2 + (n - 1) + 2}{2}$$
$$= 2 + \frac{(1^2 + 2^2 + \cdots + (n - 1)^2) + (1 + 2 + \cdots + (n - 1)) + \overbrace{2 + \cdots + 2}^{n-1}}{2}$$

parts. Since

$$1 + 2 + 3 + \cdots + (n - 1) = \frac{n(n - 1)}{2}$$

and[4]

$$1^2 + 2^2 + 3^2 + \cdots + (n-1)^2 = \frac{n(n-1)(2n-1)}{6},$$

the total number of parts is

$$2 + \frac{n(n-1)(2n-1)}{12} + \frac{n(n-1)}{4} + (n-1)$$

$$= 2 + \frac{(n-1)(2n^2 - n + 3n + 12)}{12}$$

$$= \frac{n^3 + 5n + 6}{6}.$$

45b. The solution is similar to the preceding one. For brevity we will not state the hypotheses we must impose in order to insure a maximum number of pieces but will tacitly assume them throughout. Suppose that k spheres have been drawn; let us see by how much the $(k+1)$st sphere increases the number of pieces. The $(k+1)$st sphere meets each of the first k spheres in a circle; the circles of intersection will all be different, no two of them will be tangent, and—viewed as curves on the $(k+1)$st sphere—none of them will lie inside or outside another. In problem 44b it was proved that under these conditions k circles *in a plane* will divide the plane into $k^2 - k + 2$ pieces; however, exactly the same argument can be used to prove the corresponding theorem for circles on a sphere. Therefore the surface of the $(k+1)$st sphere is divided into $k^2 - k + 2$ regions by the circles at which it intersects the first k spheres. Each of these regions splits in two one of the pieces into which the first k spheres had

---

[4] To prove this, note that:

$$2^3 = (1+1)^3 = 1^3 + 3 \cdot 1^2 + 3 \cdot 1 + 1,$$
$$3^3 = (2+1)^3 = 2^3 + 3 \cdot 2^2 + 3 \cdot 2 + 1,$$
$$4^3 = (3+1)^3 = 3^3 + 3 \cdot 3^2 + 3 \cdot 3 + 1,$$
$$\cdots\cdots\cdots\cdots\cdots\cdots\cdots\cdots\cdots\cdots$$
$$n^3 = ((n-1)+1)^3 = (n-1)^3 + 3(n-1)^2 + 3(n-1) + 1$$

Adding all these equations, we obtain

$$2^3 + 3^3 + \cdots + n^3 = 1^3 + 2^3 + \cdots + (n-1)^3$$
$$+ 3(1^2 + 2^2 + \cdots + (n-1)^2) + 3[1 + 2 + \cdots + (n-1)] + (n-1)$$

and consequently,

$$1^2 + 2^2 + \cdots + (n-1)^2 = \frac{n^3 - 1^3 - 3[1 + 2 + \cdots + (n-1)] - (n-1)}{3}$$

$$= \frac{2n^3 - 3n(n-1) - 2n}{6} = \frac{n(2n^2 - 3n + 3 - 2)}{6}$$

$$= \frac{n(n-1)(2n-1)}{6}.$$

divided space. The $(k + 1)$st sphere thus increases the number of pieces by $k^2 - k + 2$, and consequently the total number of pieces is

$$2 + (1^2 - 1 + 2) + (2^2 - 2 + 2) + (3^2 - 3 + 2)$$

$$+ \cdots + [(n - 1)^2 - (n - 1) + 2]$$

$$= 2 + (1^2 + 2^2 + 3^2 + \cdots + (n - 1)^2)$$

$$\overbrace{}^{n-1}$$

$$- (1 + 2 + 3 + \cdots + (n - 1)) + (2 + 2 + 2 + \cdots + 2)$$

$$= 2 + \frac{n(n - 1)(2n - 1)}{6} - \frac{n(n - 1)}{2} + 2(n - 1)$$

$$= 2 + \frac{(n - 1)(2n^2 - n - 3n + 12)}{6} = \frac{n(n^2 - 3n + 8)}{3}.$$

For example, five spheres can divide space into a maximum of $5(25 - 15 + 8)/3 = 30$ pieces.

46. *First solution.* Call one vertex A_1, the next one (going clockwise around the n-gon) A_2, etc. Consider a diagonal A_1A_k of the n-gon $A_1A_2A_3 \cdots A_n$. There are $k - 2$ vertices (namely, $A_2, A_3, A_4, \ldots, A_{k-1}$) on one side of this diagonal and $n - k$ vertices (namely, $A_{k+1}, A_{k+2}, \ldots, A_n$) on the other side. The diagonal A_1A_k will intersect precisely those diagonals which join one of the $k - 2$ vertices of the first group to one of the $n - k$ vertices of the second group, that is, a total of $(k - 2)(n - k)$ diagonals. Consequently, the diagonals emanating from the vertex A_1 (that is, the diagonals $A_1A_3, A_1A_4, \ldots, A_1A_{n-1}$) will intersect the other diagonals at a total of

$$1 \cdot (n - 3) + 2(n - 4) + 3(n - 5) + \cdots + (n - 3) \cdot 1$$

points. But

$$1 \cdot (n - 3) + 2(n - 4) + 3(n - 5) + \cdots + (n - 3) \cdot 1$$

$$= 1[(n - 1) - 2] + 2[(n - 1) - 3] + 3[(n - 1) - 4] + \cdots$$

$$+ (n - 3)[(n - 1) - (n - 2)]$$

$$= (n - 1)[1 + 2 + 3 + \cdots + (n - 3)]$$

$$- [1 \cdot 2 + 2 \cdot 3 + 3 \cdot 4 + \cdots + (n - 3)(n - 2)]$$

$$= (n - 1)\frac{(n - 2)(n - 3)}{2} - [1 \cdot 2 + 2 \cdot 3 + 3 \cdot 4 + \cdots$$

$$+ (n - 3)(n - 2)].$$

And since[5]

$$1 \cdot 2 + 2 \cdot 3 + 3 \cdot 4 + \cdots + (n-3)(n-2) = \frac{(n-3)(n-2)(n-1)}{3},$$

we have

$$1 \cdot (n-3) + 2(n-4) + 3(n-5) + \cdots + (n-3) \cdot 1$$

$$= \frac{(n-1)(n-2)(n-3)}{2} - \frac{(n-1)(n-2)(n-3)}{3}$$

$$= \frac{(n-1)(n-2)(n-3)}{6}. \tag{1}$$

Thus the diagonals emanating from the vertex A_1 will intersect the other diagonals in a total of $(n-1)(n-2)(n-3)/6$ points. The diagonals emanating from any other vertex will intersect the remaining diagonals in the same number of points. But in multiplying $(n-1)(n-2)(n-3)/6$ by the number of vertices of the n-gon (which equals n) we count each point of intersection four times (each of these points is the intersection of exactly two diagonals, each of which has two vertices as end-points). Thus the number of points of intersection is

$$\frac{(n-1)(n-2)(n-3)}{6} \times \frac{n}{4} = \frac{n(n-1)(n-2)(n-3)}{24}.$$

Second solution. This problem can be solved much more simply by using the formula $C_n^{\ k} = \binom{n}{k}$ (see the introduction to section I). In fact, consider any four vertices. Arrange them in clockwise order around the

---

[5] By mathematical induction it is easy to prove that

$$1 \cdot 2 \cdot 3 \cdots k + 2 \cdot 3 \cdot 4 \cdots (k+1) + 3 \cdot 4 \cdot 5 \cdots (k+2) + \cdots$$
$$+ n(n+1)(n+2) \cdots (n+k-1) = \frac{n(n+1)(n+2) \cdots (n+k)}{k+1}$$

This result also follows from the problem 57g.

We can also use the fact that

$$1 \cdot 2 + 2 \cdot 3 + 3 \cdot 4 + \cdots + n(n+1)$$
$$= 1 \cdot (1+1) + 2 \cdot (2+1) + 3 \cdot (3+1) + \cdots + n(n+1)$$
$$= (1^2 + 2^2 + 3^2 + \cdots + n^2) + (1 + 2 + 3 + \cdots + n);$$

and apply the footnote on p. 105.

Similarly,

$$1 \cdot 2 \cdot 3 + 2 \cdot 3 \cdot 4 + 3 \cdot 4 \cdot 5 + \cdots + n(n+1)(n+2)$$
$$= 1(1+1)(2+2) + 2(2+1)(2+2) + \cdots + n(n+1)(n+2)$$
$$= (1^3 + 2^3 + \cdots + n^3) + 3(1^2 + 2^2 + \cdots + n^2) + 2(1 + 2 + \cdots + n);$$

now apply the footnote on p. 105 and the remark on p. 113.

n-gon. The diagonal joining the first and third points will intersect the diagonal joining the second and fourth points, and no other pair of diagonals joining these four points will intersect within the *n*-gon (see fig. 50). By associating to each set of four vertices the point at which two of its diagonals meet, we set up a one-to-one correspondence between the points of intersection and the sets of four vertices. It follows from this that the number of points of intersection equals the number of ways one can choose four vertices from among the *n* vertices of the *n*-gon, that is, the number of combinations of *n* elements 4 at a time, which equals

$$\binom{n}{4} = \frac{n(n-1)(n-2)(n-3)}{1 \cdot 2 \cdot 3 \cdot 4}.$$

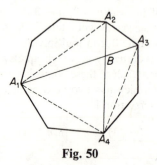

Fig. 50

47. *First solution.* Denote by f_n the number of parts into which a convex *n*-gon, no three of whose diagonals are concurrent, is divided by its diagonals. We will derive a relation connecting f_n with f_{n+1}. Consider any convex $(n + 1)$-gon; denote one of its vertices by A_1, the next vertex in the clockwise direction by A_2, the next one by A_3, etc. (fig. 51). The polygon $A_1 A_2 \cdots A_n$ is then a convex *n*-gon. Draw all of its diagonals; they will also be diagonals of the given $(n + 1)$-gon, $A_1 A_2 \cdots A_n A_{n+1}$. To obtain the rest of the diagonals of the given $(n + 1)$-gon, we must join the vertex A_{n+1} to each of the $n - 2$ nonadjacent vertices. Consider the diagonal joining A_{n+1} to A_k ($k = 2, 3, 4, \ldots, n - 1$); $k - 1$ vertices lie on one side of it (the vertices $A_1, A_2, \ldots, A_{k-1}$), and $n - k$ vertices lie on the other side (the vertices $A_{k+1}, A_{k+2}, \ldots, A_n$).

Consequently, the diagonal $A_{n+1} A_k$ meets $(k - 1)(n - k)$ of the diagonals of the $(n + 1)$-gon. The points of intersection divide this diagonal into $(k - 1)(n - k) + 1$ parts. Hence this diagonal increases by $(k - 1)(n - k) + 1$ the number of parts into which the *n*-gon is divided. The diagonals which do not pass through A_{n+1} divide the $(n + 1)$-gon into $f_n + 1$ parts. (f_n parts make up the *n*-gon $A_1 A_2 \cdots A_n$ and the

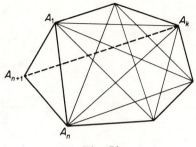

Fig. 51

$(f_n + 1)$st part is the triangle $A_{n+1}A_1A_n$). After drawing the diagonals through A_{n+1} the number of pieces is increased by

$$[(2 - 1)(n - 2) + 1] + [(3 - 1)(n - 3) + 1] + \cdots$$
$$+ [((n - 1) - 1)(n - (n - 1)) + 1].$$

We thus have

$$f_{n+1} = f_n + (2 - 1)(n - 2) + (3 - 1)(n - 3) + \cdots$$
$$+ [(n - 1) - 1][n - (n - 1)] + n - 1.$$

But

$$(2 - 1)(n - 2) + (3 - 1)(n - 3) + \cdots + [(n - 1) - 1][n - (n - 1)]$$
$$= 1 \cdot (n - 2) + 2(n - 3) + \cdots + (n - 2) \cdot 1 = \frac{(n - 2)(n - 1)n}{6},$$

by formula (1) on page 107, with n replaced by $n + 1$.

Consequently,

$$f_{n+1} = f_n + \frac{(n - 2)(n - 1)n}{6} + n - 1$$

and similarly

$$f_n = f_{n-1} + \frac{(n - 3)(n - 2)(n - 1)}{6} + n - 2,$$

$$\cdots \cdots \cdots \cdots \cdots$$

$$f_4 = f_3 + \frac{1 \cdot 2 \cdot 3}{6} + 2,$$

$$f_3 = 1.$$

Adding these equations, we obtain

$$f_{n+1} = \frac{1 \cdot 2 \cdot 3}{6} + \frac{2 \cdot 3 \cdot 4}{6} + \cdots + \frac{(n - 2)(n - 1)n}{6}$$
$$+ (1 + 2 + 3 + \cdots + (n - 1)).$$

Since[6]

$$1 \cdot 2 \cdot 3 + 2 \cdot 3 \cdot 4 + 3 \cdot 4 \cdot 5 + \cdots + (n-2)(n-1)n$$
$$= \frac{(n-2)(n-1)n(n+1)}{4},$$

we end up with

$$f_{n+1} = \frac{(n-2)(n-1)n(n+1)}{24} + \frac{n(n-1)}{2}.$$

Consequently,

$$f_n = \frac{(n-3)(n-2)(n-1)n}{24} + \frac{(n-1)(n-2)}{2}$$

$$= \frac{(n-1)(n-2)(n^2 - 3n + 12)}{24}.$$

Setting $n = 3, 4, 5, \ldots$, in this formula, we obtain:

$$f_3 = 1, \quad f_4 = 4, \quad f_5 = 11, \quad f_6 = 25, \quad f_7 = 50, \quad f_8 = 82, \ldots$$

From this result it follows in particular that if no three diagonals of a convex polygon are concurrent, then the number of pieces into which the polygon is divided by its diagonals depends only on the number of vertices and not on the shape of the polygon.

Second solution. The diagonals of an n-gon divide it into smaller polygons. We denote by r_3 the number of triangles among these polygons, by r_4 the number of quadrilaterals, by r_5 the number of pentagons, etc., and finally by r_m the number of m-gons, where m is the greatest number of sides of any of the polygons formed by the diagonals of the n-gon. We have to evaluate the sum

$$f_n = r_3 + r_4 + r_5 + \cdots + r_m.$$

Let us determine the sum of the numbers of vertices of each of the polygons into which the n-gon is divided by its diagonals. On the one hand, this sum is equal to

$$3r_3 + 4r_4 + 5r_5 + \cdots + mr_m.$$

On the other hand, each of the points of intersection of the diagonals of the n-gon is a vertex of each of the four polygons which meet there, and each of the vertices of the n-gon is a vertex of the $n - 2$ polygons which meet at that point (fig. 52). But since the number of points of intersection of the diagonals of the n-gon is $[n(n-1)(n-2)(n-3)]/24$ (see problem 44) and the number of vertices of the n-gon is n, the sum of the numbers

---

[6] See footnote on p. 107.

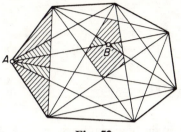

Fig. 52

of vertices of all the polygons into which the n-gon is divided by its diagonals is

$$4\frac{n(n-1)(n-2)(n-3)}{24} + (n-2)n$$

$$= \frac{n(n-1)(n-2)(n-3)}{6} + n(n-2).$$

Therefore we have

$$3r_3 + 4r_4 + 5r_5 + \cdots + mr_m = \frac{n(n-1)(n-2)(n-3)}{6} + n(n-2).$$

Now let us determine the sum of the angles of all the polygons into which the n-gon is divided. Since the sum of the angles of a k-gon is $(k-2)180°$, the required sum equals

$$[r_3 + 2r_4 + 3r_5 + \cdots + (m-2)r_m]180°.$$

On the other hand, the sum of the angles which meet at any of the $[n(n-1)(n-2)(n-3)]/24$ points of intersection of the diagonals of the n-gon is $360°$, and the sum of all the angles whose vertices are vertices of the n-gon equals the sum of the angles of the n-gon, that is, it equals $(n-2)180°$. It follows from this that

$$[r_3 + 2r_4 + 3r_5 + \cdots + (m-2)r_m]180°$$

$$= \frac{n(n-1)(n-2)(n-3)}{24}360° + (n-2)180°,$$

that is,

$$r_3 + 2r_4 + 3r_5 + \cdots + (m-2)r_m = \frac{n(n-1)(n-2)(n-3)}{12} + (n-2).$$

Subtracting the expression for $r_3 + 2r_4 + 3r_5 + \cdots + (m-2)r_m$ from the expression for $3r_3 + 4r_4 + 5r_5 + \cdots + mr_m$, we obtain:

$$2(r_3 + r_4 + r_5 + \cdots + r_m)$$

$$= \frac{n(n-1)(n-2)(n-3)}{12} + (n-1)(n-2),$$

whence

$$f_n = r_3 + r_4 + r_5 + \cdots + r_m$$
$$= \frac{n(n-1)(n-2)(n-3)}{24} + \frac{(n-1)(n-2)}{2}$$
$$= \frac{(n-1)(n-2)(n^2-3n+12)}{24}.$$

48a. Let us first ascertain how many rectangles k squares wide and l squares high can be formed on a chessboard. Every such rectangle is obtained by selecting k consecutive columns and l consecutive rows (fig. 53). The group of k consecutive columns can be chosen in $9 - k$

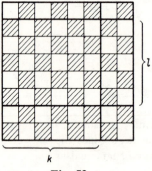

Fig. 53

different ways (the last column of the group can be the k-th, $(k + 1)$st, ..., or 8th column of the board); similarly, the group of l rows can be chosen in $9 - l$ ways. It follows from this that a rectangle k squares wide and l squares high can be chosen in $(9 - k)(9 - l)$ different ways, and that there are a total of $(9 - k)(9 - l)$ such rectangles on the board.

Let us determine how many rectangles of given width k there are on the board. The height l of the rectangle ranges from 1 to 8; hence the number of such rectangles is

$$(9 - k)(9 - 1) + (9 - k)(9 - 2) + \cdots + (9 - k)(9 - 8)$$
$$= (9 - k)(8 + 7 + 6 + 5 + 4 + 3 + 2 + 1) = 36(9 - k).$$

Now taking into account that the width k of the rectangle also ranges from 1 to 8, we find that the total number of different rectangles is

$$36(9 - 1) + 36(9 - 2) + \cdots + 36(9 - 8)$$
$$= 36(8 + 7 + 6 + 5 + 4 + 3 + 2 + 1) = 36 \cdot 36 = 1296.$$

48b. As in the solution of part a, we conclude that the total number of rectangles k squares wide and l squares high on an $n \times n$ board is $(n + 1 - k)(n + 1 - l)$. The total number of rectangles of width k is therefore

$$(n + 1 - k)(1 + 2 + \cdots + n) = (n + 1 - k) \frac{n(n + 1)}{2}.$$

Hence the number of all rectangles on the board is

$$(1 + 2 + \cdots + n) \frac{n(n + 1)}{2} = \left[\frac{n(n + 1)}{2} \right]^2.$$

49a. This problem is closely related to the preceding one. The number of different $k \times k$ sections of the board is $(9 - k)^2$ (see solution to problem 48a). It follows from this that the total number of square sections is

$$8^2 + 7^2 + 6^2 + \cdots + 1^2$$
$$= 64 + 49 + 36 + 25 + 16 + 9 + 4 + 1 = 204.$$

49b. By an argument similar to that of part a, we conclude that the required number is

$$1^2 + 2^2 + 3^2 + \cdots + n^2 = \frac{n(n + 1)(2n + 1)}{6}$$

(see, for example, the formula given in the footnote on p. 105; the above formula can be obtained from that one by substituting $n + 1$ for n).

Remark. Since by a well-known formula (which can be derived without difficulty by mathematical induction or by the method outlined in the footnote on p. 105), we have

$$1^3 + 2^3 + 3^3 + \cdots + n^3 = \left(\frac{n(n + 1)}{2} \right)^2,$$

the results of parts a and b can be given in the following symmetric form:

The number of different square sections on an $n \times n$ board is

$$1^2 + 2^2 + 3^2 + \cdots + n^2;$$

the number of different rectangular sections is

$$1^3 + 2^3 + 3^3 + \cdots + n^3.$$

50. It is obvious that the vertices of the triangles in question must be either vertices of the n-gon or intersection points of its diagonals. Let us consider separately the following four cases:

    1. All three vertices of the triangle are vertices of the n-gon.

    2. Two vertices of the triangle are vertices of the n-gon and the third vertex is a point where two diagonals intersect.

    3. One vertex of the triangle is a vertex of the n-gon and the other two are intersection points of diagonals.

4. All three vertices of the triangle are intersection points of diagonals.

Case 1.   The number of triangles all of whose vertices are vertices of the *n*-gon (fig. 54a) is $C_n{}^3 = \binom{n}{3}$.

Case 2.   Consider any triangle A_1A_2B, where A_1 and A_2 are vertices of the *n*-gon and B is a point where two diagonals intersect (fig. 54b). The sides A_1B and A_2B of this triangle are parts of diagonals A_1A_3 and A_2A_4 of the *n*-gon; our triangle is one of the four triangles into which the quadrilateral $A_1A_2A_3A_4$ is divided by its diagonals. Thus each

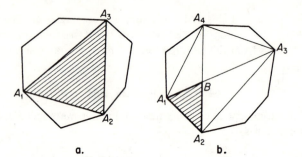

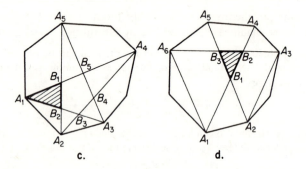

Fig. 54

(unordered) quadruple A_1, A_2, A_3, A_4 of vertices of the original *n*-gon accounts for four triangles of type 2.

Consequently, the total number of triangles of type 2 is $4\binom{n}{4}$.

Case 3.   Consider any triangle $A_1B_1B_2$, where A_1 is a vertex of the *n*-gon and B_1 and B_2 are points of intersection of the diagonals (fig. 54c). The sides A_1B_1, A_1B_2, and B_1B_2 of this triangle are contained in diagonals A_1A_4, A_1A_3, and A_2A_5 of the *n*-gon; our triangle $A_1B_1B_2$ is one of the five triangles formed by the diagonals of the pentagon $A_1A_2A_3A_4A_5$ (that is, it is one of the five triangular "points" of the star $A_1B_2A_2B_3A_3B_4A_4B_5A_5B_1$). Thus each quintuple of vertices A_1, A_2, A_3, A_4, A_5 accounts for five triangles of type 3. Consequently, the total number of triangles of type 3 is $5\binom{n}{5}$.

Case 4. Consider any triangle $B_1B_2B_3$, where B_1, B_2, and B_3 are intersection points of diagonals of the n-gon (fig. 54d). The sides B_1B_2, B_2B_3, and B_3B_1 of this triangle are contained in diagonals A_1A_4, A_2A_5, and A_3A_6 of the n-gon. Each sextuple of vertices A_1, A_2, A_3, A_4, A_5, A_6 of our n-gon accounts for the single triangle of type 4 formed by the diagonals joining opposite vertices of the hexagon $A_1A_2A_3A_4A_5A_6$. Consequently, the total number of triangles of type 4 is $\binom{n}{6}$.

Combining the four cases, we find that the total number T_n of triangles formed by the edges and diagonals of the convex n-gon is

$$\begin{aligned}
T_n &= \binom{n}{3} + 4\binom{n}{4} + 5\binom{n}{5} + \binom{n}{6} \\
&= \frac{n(n-1)(n-2)}{6}\left[1 + (n-3) + \frac{(n-3)(n-4)}{4}\right. \\
&\quad \left. + \frac{(n-3)(n-4)(n-5)}{120}\right] \\
&= \frac{n(n-1)(n-2)(n^3 + 18n^2 + 43n + 60)}{720}.
\end{aligned}$$

Setting $n = 3, 4, 5$, etc., in this formula, we get

$$T_3 = 1, \ T_4 = 8, \ T_5 = 35, \ T_6 = 111, \text{ etc.}$$

51. In order that there be any such k-gons, it is necessary that n be at least $2k$ (since any two vertices of the k-gon must have at least one other vertex of the n-gon between them).

Denote the vertices of the n-gon by the letters A_1, A_2, $\ldots$, A_{n-1}, A_n (the subscripts increase from 1 to n as one goes counterclockwise around the n-gon); let us compute how many k-gons there are which satisfy the hypothesis of the problem and have A_{n-1} as a vertex. Let the remaining $k - 1$ vertices of any such k-gon be (in order of increasing subscripts) $A_{i_1}, A_{i_2}, \ldots, A_{i_{k-1}}$. The numbers $i_1, i_2, \ldots, i_{k-1}$ lie between 1 and $n - 3$ (inclusive) and satisfy the condition

$$i_2 - i_1 \geqq 2, i_3 - i_2 \geqq 2, \ldots, i_{k-1} - i_{k-2} \geqq 2.$$

Now consider the $k - 1$ numbers

$$j_1 = i_1, j_2 = i_2 - 1, j_3 = i_3 - 2, \ldots, j_{k-1} = i_{k-1} - (k - 2).$$

From the inequalities which the numbers $i_1, i_2, \ldots, i_{k-1}$ must satisfy, it follows that $j_1, j_2, \ldots, j_{k-1}$ satisfy the inequalities

$$1 \leqq j_1 < j_2 < j_3 < \cdots < j_{k-1} \leqq (n-3) - (k-2) = n-k-1.$$

Conversely, if $j_1, j_2, \ldots, j_{k-1}$ are distinct integers between 1 and $n-k-1$ arranged in increasing order, then the numbers $i_1 = j_1, i_2 = j_2 + 1, \ldots, i_{k-1} = j_{k-1} + (k-2)$ will satisfy the inequalities

$$i_1 \geqq 1, i_2 - i_1 \geqq 2, \ldots, i_{k-1} - i_{k-2} \geqq 2, i_{k-1} \leqq n-3$$

and consequently, the k-gon $A_{i_1} A_{i_2} \cdots A_{i_{k-1}} A_{n-1}$ will be of the type under consideration. It follows from this that the number of such k-gons which have the point A_{n-1} as a vertex is equal to the number of ways one can choose $k-1$ distinct positive integers which do not exceed $n-k-1$, that is, it equals $\binom{n-k-1}{k-1}$.

It is now easy to compute how many k-gons satisfy the conditions of the problem. Since the number of k-gons having a given vertex of the

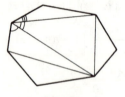

Fig. 55

n-gon among their vertices is the same for any vertex, by multiplying $\binom{n-k-1}{k-1}$ by n (that is, by adding up the number of k-gons having A_1 as a vertex, the number having A_2 as a vertex, ... and the number having A_n), we count each k-gon which satisfies the hypotheses of the problem k times (since each k-gon has k vertices). Consequently, the total number of such k-gons is

$$\frac{n}{k}\binom{n-k-1}{k-1} = \frac{n(n-k-1)!}{k!\,(n-2k)!}.$$

52a. Let the n-gon be split into k triangles by diagonals which do not intersect inside it. Then the sum of the interior angles of all these triangles is $k \cdot 180°$. Let us now evaluate this sum in a different way. Since our diagonals do not intersect within the n-gon, the vertices of the triangles must all be vertices of the n-gon (fig. 55). The sum of the angles of the triangles is therefore equal to the sum of the angles of the n-gon, which is $(n-2) \cdot 180°$. It follows from this that $k \cdot 180° = (n-2) \cdot 180°$, or

$$k = n - 2.$$

Hence the number of triangles into which an n-gon is divided by the diagonals is always $n - 2$, and does not depend on the way the n-gon is divided.

52b. Let us now compute l, the number of diagonals involved in such a decomposition. Each triangle has three sides; consequently, the total number of sides of all the triangles in such a decomposition is $3 \cdot (n - 2)$. But each diagonal involved is a side of two of the triangles, and each side of the n-gon is a side of one of the triangles. Consequently,

$$3(n - 2) = 2l + n.$$

Hence

$$3n - 6 - n = 2l, \qquad l = \frac{2n - 6}{2} = n - 3.$$

Therefore, the number of diagonals is $n - 3$ and is independent of the way the n-gon is divided.

53a. Denote by T_n the number of ways in which the given n-gon P can be divided into triangles by diagonals which do not intersect inside P. It is convenient to make the convention that $T_2 = 1$. We will first derive an expression for T_n in terms of $T_2, T_3, T_4, \ldots, T_{n-1}$.

Select any side of P, say $A_1 A_2$. When P is decomposed into triangles, this side must occur in one of them, and the remaining vertex of that triangle is some A_k, where $3 \leq k \leq n$. For a fixed value of k let us compute how many such decompositions there are. The diagonal $A_2 A_k$ cuts off the $(k - 1)$gon $A_2 A_3 \cdots A_k$ from P. This can be divided into triangles in T_{k-1} ways. Similarly the diagonal $A_1 A_k$ cuts off the $(n - k + 2)$gon $A_k A_{k+1} \cdots A_n A_1$, and this can be divided into triangles in T_{n-k+2} ways. Therefore the number of decompositions of P in which the triangle $A_1 A_2 A_k$ occurs is $T_{k-1} T_{n-k+2}$. Assigning to k all values from 3 to n and adding, we obtain the relation

$$T_n = T_2 T_{n-1} + T_3 T_{n-2} + T_4 T_{n-3} + \cdots + T_{n-2} T_3 + T_{n-1} T_2. \qquad (1)$$

Thus we have

$$T_3 = T_2 \cdot T_2 = 1$$
$$T_4 = T_2 \cdot T_3 + T_3 \cdot T_2 = 2$$
$$T_5 = T_2 \cdot T_4 + T_3 \cdot T_3 + T_4 \cdot T_2 = 5$$
$$T_6 = T_2 \cdot T_5 + T_3 \cdot T_4 + T_4 \cdot T_3 + T_5 \cdot T_2 = 14$$
$$T_7 = T_2 \cdot T_6 + T_3 \cdot T_5 + T_4 \cdot T_4 + T_5 \cdot T_3 + T_6 \cdot T_2 = 42$$
$$T_8 = T_2 \cdot T_7 + T_3 \cdot T_6 + T_4 \cdot T_5 + T_5 \cdot T_4 + T_6 \cdot T_3 + T_7 \cdot T_2 = 132.$$

Therefore a convex octagon can be divided into triangles in 132 different ways by diagonals which do not intersect inside the octagon.

53b. Let us begin by rewriting formula (1) of part a with n replaced by $n + 1$:

$$T_{n+1} = T_2 T_n + T_3 T_{n-1} + T_4 T_{n-2} + \cdots + T_{n-1} T_3 + T_n T_2$$

or

$$T_{n+1} - 2T_n = T_3 T_{n-1} + T_4 T_{n-2} + \cdots + T_{n-1} T_3. \tag{2}$$

In order to get an explicit expression for T_n, we will derive another recursion formula for T_n in terms of $T_3, \ldots, T_{n-1}$ which will then be combined with (2).

Let us compute the number of decompositions of P in which the diagonal $A_1 A_k$ is involved. This diagonal divides P into the k-gon $A_1 A_2 \cdots A_k$ and the $(n - k + 2)$gon $A_k A_{k+1} \cdots A_n A_1$ (fig. 56). Since the first of these two polygons can be divided into triangles in T_k ways and

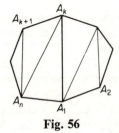

Fig. 56

the second in T_{n-k+2} ways, the number of decompositions of P in which the diagonal $A_1 A_k$ is involved equals $T_k T_{n-k+2}$.

Let us add up the numbers of decompositions involving each of the diagonals of P. The sum corresponding to the $n - 2$ diagonals emanating from A_1 is

$$T_3 T_{n-1} + T_4 T_{n-2} + \cdots + T_{n-2} T_4 + T_{n-1} T_3. \tag{3}$$

The sum corresponding to any of the other vertices will have the same value. Each diagonal figures in two of these sums (namely, the sums corresponding to its two endpoints). Consequently, n times the sum (3) is twice the sum of the numbers obtained by counting the number of decompositions in which each diagonal occurs.

According to the result of problem 52b, the number of diagonals involved in a decomposition of a convex n-gon into triangles is $n - 3$. Thus, in the expression

$$\frac{n}{2} (T_3 T_{n-1} + T_4 T_{n-2} + \cdots + T_{n-2} T_4 + T_{n-1} T_3)$$

each decomposition is counted $n - 3$ times: once for each of the $n - 3$ diagonals involved in it. Consequently,

$$\frac{n}{2}(T_3 T_{n-1} + T_4 T_{n-2} + \cdots + T_{n-2} T_4 + T_{n-1} T_3) = (n - 3) T_n. \quad (4)$$

This is the new recursion formula which we wanted to obtain.

By equation (2), we can replace the parenthesis on the left of (4) by $T_{n+1} - 2T_n$. This gives

$$\frac{n}{2}(T_{n+1} - 2T_n) = (n - 3) T_n,$$

whence

$$T_{n+1} - 2T_n = \frac{2(n - 3)}{n} T_n,$$

$$T_{n+1} = 2T_n + \frac{2(n - 3)}{n} T_n = \frac{4n - 6}{n} T_n = \frac{2(2n - 3)}{n} T_n.$$

Thus we see that

$$T_3 = 1$$

$$T_4 = \frac{2 \cdot 3}{3} T_3 = \frac{2^2 \cdot 3}{2 \cdot 3}$$

$$T_5 = \frac{2 \cdot 5}{4} T_4 = \frac{2^3 \cdot 3 \cdot 5}{2 \cdot 3 \cdot 4}$$

$$T_6 = \frac{2 \cdot 7}{5} T_5 = \frac{2^4 \cdot 3 \cdot 5 \cdot 7}{2 \cdot 3 \cdot 4 \cdot 5},$$

and in general

$$T_n = \frac{2^{n-2} \cdot 3 \cdot 5 \cdot 7 \cdots (2n - 5)}{(n - 1)!}$$

Using binomial coefficients, this expression can be rewritten in the following form:

$$T_n = \frac{1 \cdot 2 \cdot 3 \cdot 4 \cdots (2n - 6)(2n - 5)}{(n - 1)! \, 2 \cdot 4 \cdots (2n - 6)} 2^{n-2} = \frac{(2n - 5)! \, 2^{n-2}}{(n - 1)! \, 2^{n-3}(n - 3)!}$$

$$= \frac{2}{n - 1} \frac{(2n - 5)!}{(n - 2)! \, (n - 3)!} = \frac{2}{n - 1} \binom{2n - 5}{n - 3}$$

or in another form,

$$T_n = \frac{(2n - 5)! \, 2}{(n - 1)! \, (n - 3)!} = \frac{(2n - 4)!}{(2n - 2)(n - 1)! \, (n - 3)!} = \frac{(2n - 4)!}{(n - 1)! \, (n - 2)!}$$

$$= \frac{1}{n - 1} \frac{(2n - 4)!}{[(n - 2)!]^2} = \frac{1}{n - 1} \binom{2n - 4}{n - 2}$$

or,

$$T_n = \frac{(2n-4)!}{(n-1)!\,(n-2)!}\,\frac{1}{2n-3}\,\frac{(2n-3)!}{(n-1)!\,(n-2)!} = \frac{1}{2n-3}\binom{2n-3}{n-2}.$$

For large values of n, the computation of T_n by any of the above formulas is quite lengthy. However, by using Stirling's formula[7]

$$n! \sim \sqrt{2\pi n}\left(\frac{n}{e}\right)^n$$

one can obtain without difficulty an approximate formula for T_n:

$$T_n = \frac{(2n-4)!}{(n-1)[(n-2)!]^2} \sim \frac{\sqrt{2\pi 2(n-2)}\,[2(n-2)]^{2(n-2)}e^{-2(n-2)}}{(n-1)[\sqrt{2\pi(n-2)}(n-2)^{n-2}e^{-(n-2)}]^2}$$
$$= \frac{2^{2(n-2)}}{(n-1)\sqrt{\pi(n-2)}}.$$

This formula allows us to estimate T_n with the aid of a logarithm table for large values of n. The relative error approaches zero as $n \to \infty$.

54. Let F_n be the number of ways in which $2n$ points on the circumference of a circle can be joined in pairs by n chords which do not intersect within the circle. We shall first obtain an expression for F_n in terms of $F_1, F_2, \ldots, F_{n-1}$.

Denote one of the points by A_1, and then, proceeding around the circle in a counterclockwise direction, let the remaining points be $A_2, A_3, \ldots, A_{2n}$. The point A_1 can be connected to any of the points $A_2, A_4, A_6, \ldots, A_{2n}$, but to no others; for if it were connected to a point A_m with m odd, there would be an odd number of points on each side of the chord $A_1 A_m$, so that no matter how these points were joined in pairs, one of the chords would have to cross $A_1 A_m$.

Let us now compute how many ways of joining the points are such that A_1 is joined to A_{2k}. On one side of the chord $A_1 A_{2k}$ there are $2(k-1)$ points (the points $A_2, A_3, \ldots, A_{2k-1}$), and on the other side there are $2(n-k)$ points (the points $A_{2k+1}, A_{2k+2}, \ldots, A_{2n}$). The first $2(k-1)$ points can be joined in F_{k-1} ways and the other $2(n-k)$ points in F_{n-k} ways (here we are making the convention that $F_0 = 1$).

Therefore the number of solutions in which A_1 is joined to A_{2k} is $F_{k-1}F_{n-k}$. Letting k run from 1 to n and adding, we get

$$F_n = F_0F_{n-1} + F_1F_{n-2} + F_2F_{n-3} + \cdots + F_{n-2}F_1 + F_{n-1}F_0.$$

[7] See for example R. Courant, *Differential and Integral Calculus*, New York Interscience, 1937, p. 361–364.

This formula is very similar to the one that was obtained for T_n in the previous problem. Let us first replace n by $n + 2$ in equation (1) of problem 53a. This gives

$$T_{n+2} = T_2 T_{n+1} + T_3 T_n + T_4 T_{n-1} + \cdots + T_n T_3 + T_{n+1} T_2.$$

If we put $G_n = T_{n+2}$ this becomes

$$G_n = G_0 G_{n-1} + G_1 G_{n-2} + G_2 G_{n-3} + \cdots + G_{n-2} G_1 + G_{n-1} G_0.$$

Thus F_n and G_n satisfy the same recursion formula. Moreover $F_0 = 1$ and $G_0 = T_2 = 1$. Consequently, $F_n = G_n$ for all values of n, and so

$$F_n = T_{n+2} = \frac{1 \cdot 3 \cdot 5 \cdots (2n - 1)}{(n + 1)!} 2^n.$$

Using the expressions we derived for T_n in terms of binomial coefficients,

$$F_n = \frac{2}{n + 1} \binom{2n - 1}{n - 1} = \frac{1}{n + 1} \binom{2n}{n} = \frac{1}{2n + 1} \binom{2n + 1}{n}.$$

55a. Denote one of the sectors by S_0, and then, proceeding around the circle in a counterclockwise direction, denote the other sectors by S_1, S_2, ..., S_{p-1}. Put $S_p = S_0$, $S_{p+1} = S_1, \ldots, S_{2p} = S_0$, $S_{2p+1} = S_1$, etc. Suppose for the moment that we consider two colorings C_1 and C_2 of the circle to be different if they differ in the way any one of the sectors is colored. Since each sector can be colored in n ways, there are n^p colorings of the circle. This is not the answer to the problem, however, because some of these colorings can be obtained from others by rotating the circle. Let us write $C_1 \sim C_2$ (read: C_1 is equivalent to C_2) if C_1 can be obtained from C_2 by rotating the circle.

We can now divide the set of all colorings up into classes, by putting two colorings C_1 and C_2 in the same class whenever $C_1 \sim C_2$. Thus, the class containing a given coloring C_1 consists of all colorings equivalent to C_1; we call it the *equivalence class* of C_1.[8]

---

[8] The relation $C_1 \sim C_2$ has the following three properties

    (a) $C_1 \sim C_1$
    (b) If $C_1 \sim C_2$, then $C_2 \sim C_1$
    (c) If $C_1 \sim C_2$ and $C_2 \sim C_3$, then $C_1 \sim C_3$.

A relation having these three properties is called an *equivalence relation*. A set in which an equivalence relation is defined can always be divided up into equivalence classes by putting C_1 and C_2 in the same class whenever $C_1 \sim C_2$. For a more complete description of this procedure, see, e.g., R. E. Johnson, *First Course in Abstract Algebra*, Prentice-Hall, 1953, pp. 8–11, or van der Waerden, *Modern Algebra*, Vol. I, Ungar, 1953, pp. 9–10.

The problem is to determine the number of equivalence classes. To do this it is essential to know how many colorings are equivalent to a given coloring C. First of all if C is a coloring in which all p sectors are painted the same color, then C is equivalent only to itself. Therefore we have n equivalence classes (one for each of the n colors) consisting of a single coloring. If C_0 is a coloring in which at least two sectors have different colors, then we will show that the colorings $C_0, C_1, C_2, \ldots, C_{p-1}$ obtained by rotating C_0 counterclockwise through angles of $0°$, $360°/p$, $2 \cdot (360°/p), \ldots, (p-1)(360°/p)$, are all different. (Note that if p is not prime this need not be true; for example, by rotating the coloring illustrated in fig. 57 through an angle of $2 \cdot (360°/6) = 120°$, we obtain the same coloring.)

Suppose then that $C_i = C_j$, where $0 \leqq i < j \leqq p - 1$. Putting $k = j - i$, this means that the coloring C_i is not changed by a rotation through

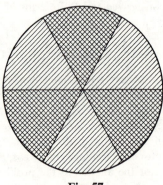

Fig. 57

$k \cdot (360°/p)$, where $0 < k < p$. Say for the sake of the argument that S_0 is colored red in the coloring C_i. Then S_k must also be red, and the same is true of S_{2k}, S_{3k}, etc. Now the sectors $S_0, S_k, S_{2k}, \ldots, S_{(p-1)k}$ are all different. For if $S_{lk} = S_{mk}$ where $0 \leqq l < m \leqq p - 1$, then $mk - lk = (m - l)k$ would be a multiple of p. This is impossible since $0 < m - l < p$, $0 < k < p$, and p is prime. Since there are only p sectors altogether, every sector is therefore of the form S_{lk}, and so is painted red in the coloring C_i. This contradicts the fact that not all sectors were painted the same color in the coloring C_0 (for C_i is a rotation of C_0).

Thus we have shown that when C_i is a coloring in which two sectors have different colors, the equivalence class of C_0 contains exactly p members (namely $C_0, C_1, \ldots, C_{p-1}$).

Let N denote the total number of equivalence classes. Since we have seen that there are n equivalence classes of one member each, there must be $N - n$ equivalence classes of p members each. Since the total number of colorings is n^p, we therefore have

$$n + (N - n)p = n^p$$

or solving for N,

$$N = \frac{n^p - n}{p} + n.$$

55b. Since the number N of part a is an integer by its definition, it follows that for any n the number $n^p - n$ is divisible by p. This is Fermat's theorem.

56a. Let the points be $A_0, A_1, A_2, \ldots, A_{p-1}$ going around the circle in the counterclockwise direction. Put $A_p = A_0$, $A_{p+1} = A_1, \ldots, A_{2p} = A_0$, $A_{2p+1} = A_1$, etc. Let us first compute how many self-intersecting polygons there are with vertices at $A_0, A_1, \ldots, A_{p-1}$, counting two polygons as different if they differ either in shape or location. To obtain a polygon we join A_0 to any point A_{i_1} other than A_0, then join A_{i_1} to any point A_{i_2} other than A_0 and A_{i_1}, and continue in this way until all the points are exhausted; then we join the last point $A_{i_{p-1}}$ to A_0. The point A_{i_1} can be chosen in $p - 1$ ways; once it is chosen, A_{i_2} can be chosen in $p - 2$ ways, etc. Therefore the total number of ways of choosing the sequence $A_{i_1}, A_{i_2}, \ldots,$ $A_{i_{p-1}}$ is $(p - 1)(p - 2)(p - 3) \cdots 1 = (p - 1)!$ Note, however, that each polygon having $A_0, A_1, \ldots, A_{p-1}$ as its vertices is obtained twice by this process, once in the form $A_0 A_{i_1} A_{i_2} \cdots A_{i_{p-1}} A_0$ and once in the form $A_0 A_{i_{p-1}} A_{i_{p-2}} \cdots A_{i_1} A_0$. (For example, the heptagons $A_0 A_3 A_4 A_2 A_6 A_1 A_5 A_0$ and $A_0 A_5 A_1 A_6 A_2 A_4 A_3 A_0$ coincide but differ from all other heptagons $A_0, A_{i_1} A_{i_2} \cdots A_{i_6} A_0$.)

Thus the total number of p-gons with the given points as vertices is $(p - 1)!/2$. Among these p-gons exactly one is not self-intersecting, namely $A_0 A_1 A_2 \cdots A_{p-1} A_0$. The others have at least one pair of sides which cross at an interior point. Therefore the number of self-intersecting polygons is $(p - 1)!/2 - 1$. This is not the answer to the problem because some of these polygons can be obtained from others by rotating the circle. As in problem 55 we will say that two polygons P and Q are equivalent if they can be obtained from each other by rotating the circle; we then write $P \sim Q$.

The set of all self-intersecting polygons is now broken up into classes by putting P and Q in the same class whenever $P \sim Q$. The class in which a polygon P lies consists of all polygons equivalent to P; it is called the *equivalence class* of P. Our problem is to determine the number of equivalence classes.

Let P_0 be any polygon, and denote by $P_0, P_1, P_2, \ldots, P_{p-1}$ the polygons obtained by rotating P_0 counterclockwise through angles of $0°$, $360°/p$, $2(360°/p), \ldots, (p - 1)(360°/p)$. We will prove that $P_0, P_1, \ldots,$ P_{p-1} are either all different or all the same.

Suppose they are not all different, so that $P_i = P_j$, where $0 \leq i < j \leq p - 1$. Putting $k = j - i$, it follows that P_i is unchanged when the circle is rotated through $k \cdot (360°/p)$, where $0 < k < p$. This means that if A_0 is joined to some point A_t in the polygon P_i, then A_k must be joined to A_{k+t}, A_{2k} must be joined to A_{2k+t}, etc. As in problem 55 we can show that the points $A_0, A_k, A_{2k}, \ldots, A_{(p-1)k}$ are all different. For if $A_{lk} = A_{mk}$, where $0 \leq l < m \leq p - 1$, then $mk - lk = (m - l)k$ would be a multiple of p, which is impossible, since $0 < m - l < p$, $0 < k < p$, and p is prime. Since there are only p vertices altogether, $A_0, A_k, \ldots, A_{(p-1)k}$ therefore constitute all the vertices. Thus every vertex A_s is joined to A_{s+t}, so that the polygon P_i is regular (i.e., all its sides are equal and all its vertex angles are equal). In this case it is clear that

$$P_0 = P_1 = \cdots = P_{p-1}.$$

We have now shown that the equivalence class of a non-regular polygon has p members, while that of a regular polygon has only one member. The next step is to determine how many regular self-intersecting p-gons there are. We saw that in a regular polygon each vertex A_s is joined to A_{s+t}, where t is a fixed number satisfying $1 \leq t \leq p - 1$. But as in our discussion of the method for obtaining all polygons, each regular polygon will arise twice in this process, since t and $p - t$ give rise to the same polygon ($t \neq p - t$ since p is odd). Therefore there are $(p - 1)/2$ regular polygons; since exactly one of these non-self-intersecting, there are $(p - 1)/2 - 1 = (p - 3)/2$ self-intersecting regular polygons.

Now let N denote the total number of equivalence classes of self-intersecting polygons. Since there are $(p - 3)/2$ of these classes with one member, there are $N - (p - 3)/2$ of them with p members. Therefore, recalling that the total number of self-intersecting polygons is $(p - 1)!/2 - 1$, we get

$$\frac{p - 3}{2} + \left(N - \frac{p - 3}{2} \right)p = \frac{(p - 1)!}{2} - 1.$$

Solving for N,

$$N = \frac{1}{2} \left\{ \frac{(p - 1)! + 1}{p} + p - 4 \right\}$$

56b. When $p = 2$, we have $[(p - 1)! + 1]/p = (1 + 1)/2 = 1$.

For $p > 2$, note that in part a N is an integer by its definition. Hence $2N$ is also an integer, which implies that $[(p - 1)! + 1]/p$ is an integer. This means that $(p - 1)! + 1$ is divisible by p.

Note that if p is not prime, then $(p - 1)! + 1$ cannot be divisible by p. For in this case p has a divisor d with $1 < d < p$. Since $(p - 1)!$ is divisible by d, $(p - 1)! + 1$ is not divisible by d, and so is not divisible by p.

V. PROBLEMS ON THE BINOMIAL COEFFICIENTS

57a. $\binom{n}{0} + \binom{n}{1} + \binom{n}{2} + \cdots + \binom{n}{n} = (1+1)^n = 2^n$.

57b. $\binom{n}{0} - \binom{n}{1} + \binom{n}{2} - \cdots + (-1)^n \binom{n}{n} = (1-1)^n = 0$ if $n > 0$.

When $n = 0$, the sum in question reduces to the single term $\binom{0}{0} = 1$.

57c. We make use of the fact that

$$\frac{n+1}{k+1}\binom{n}{k} = \frac{(n+1)n(n-1)\cdots(n-k+1)}{1 \cdot 2 \cdot 3 \cdots k(k+1)} = \binom{n+1}{k+1}.$$

It follows from this that

$$(n+1)\left\{ \binom{n}{0} + \frac{1}{2}\binom{n}{1} + \frac{1}{3}\binom{n}{2} + \cdots + \frac{1}{n+1}\binom{n}{n} \right\}$$

$$= \binom{n+1}{1} + \binom{n+1}{2} + \binom{n+1}{3} + \cdots + \binom{n+1}{n+1}$$

$$= 2^{n+1} - 1,$$

so that

$$\binom{n}{0} + \frac{1}{2}\binom{n}{1} + \frac{1}{3}\binom{n}{2} + \cdots + \frac{1}{n+1}\binom{n}{n} = \frac{2^{n+1} - 1}{n+1}.$$

57d. We make use of the fact that if $n \geq 1$,

$$k\binom{n}{k} = n\frac{(n-1)(n-2)\cdots(n-k+1)}{1 \cdot 2 \cdots (k-1)} = n\binom{n-1}{k-1}.$$

It follows that when $n \geq 1$,

$$\binom{n}{1} + 2\binom{n}{2} + 3\binom{n}{3} + \cdots + n\binom{n}{n}$$

$$= n\left\{ \binom{n-1}{0} + \binom{n-1}{1} + \binom{n-1}{2} + \cdots + \binom{n-1}{n-1} \right\} = n \cdot 2^{n-1}.$$

57e. Proceeding as in part d we obtain

$$\binom{n}{1} - 2\binom{n}{2} + 3\binom{n}{3} - \cdots + (-1)^{n-1}n\binom{n}{n}$$

$$= n\left\{ \binom{n-1}{0} - \binom{n-1}{1} + \binom{n-1}{2} - \cdots + (-1)^{n-1}\binom{n-1}{n-1} \right\}$$

$$= 0 \quad \text{if } n > 1.$$

If $n = 1$ the expression to be evaluated reduces to the single term $\binom{1}{1} = 1$.

57f. The required sum is the coefficient of x^n in the polynomial

$$x^n(1 - x)^n + x^{n-1}(1 - x)^n + \cdots + x^{n-m}(1 - x)^n.$$

Transforming this polynomial, we obtain:

$$x^n(1 - x)^n + x^{n-1}(1 - x)^n + \cdots + x^{n-m}(1 - x)^n$$
$$= (1 - x)^n(x^n + x^{n-1} + x^{n-2} + \cdots + x^{n-m})$$
$$= (1 - x)^n \frac{x^{n+1} - x^{n-m}}{x - 1} = -(1 - x)^{n-1}\{x^{n+1} - x^{n-m}\}$$
$$= x^{n-m}(1 - x)^{n-1} - x^{n+1}(1 - x)^{n-1}.$$

Consequently, the coefficient of x^n is $(-1)^m \binom{n-1}{m}$ for $m < n$ and 0 for $m = n$ (compare with part b).

57g. The sum to be evaluated equals the coefficient of x^k in the polynomial

$$(1 + x)^n + (1 + x)^{n+1} + (1 + x)^{n+2} + \cdots + (1 + x)^{n+m}.$$

Making use of the formula for the sum of a geometric progression, we obtain:

$$(1 + x)^n + (1 + x)^{n+1} + \cdots + (1 + x)^{n+m}$$
$$= \frac{(1 + x)^{n+m+1} - (1 + x)^n}{(1 + x) - 1} = \frac{1}{x}\{(1 + x)^{n+m+1} - (1 + x)^n\}.$$

It is clear from this that the coefficient of x^k is $\binom{n + m + 1}{k + 1} - \binom{n}{k + 1}$ for $k < n$ and $\binom{n + m + 1}{n + 1}$ for $k = n$.

57h. The expression to be evaluated equals the coefficient of x^{2n} in the polynomial

$$x^{2n}(1 - x)^{2n} + x^{2n-1}(1 - x)^{2n-1} + x^{2n-2}(1 - x)^{2n-2} + \cdots + x^n(1 - x)^n.$$

First of all we add to this polynomial the following terms, which are of degree less than $2n$:

$$x^{n-1}(1 - x)^{n-1} + x^{n-2}(1 - x)^{n-2} + \cdots + x(1 - x) + 1.$$

Of course, adding on these terms does not affect the coefficient of x^{2n}. The sum obtained is

$$(x - x^2)^{2n} + (x - x^2)^{2n-1} + (x - x^2)^{2n-2} + \cdots + (x - x^2) + 1$$
$$= \frac{(x - x^2)^{2n+1} - 1}{(x - x^2) - 1} = \frac{x^{2n+1}(1 - x)^{2n+1} - 1}{-x^2 + x - 1} = \frac{1 - x^{2n+1}(1 - x)^{2n+1}}{x^2 - x + 1}.$$

Consequently, we have to compute the coefficient of x^{2n} in the expression

$$[1 - x^{2n+1}(1 - x)^{2n+1}] \frac{1}{1 - x + x^2} = [1 - x^{2n+1}(1 - x)^{2n+1}] \frac{1 + x}{1 + x^3}.$$

By virtue of the formula for the sum of an infinite geometric progression,

$$\frac{1 + x}{1 + x^3} = (1 + x) \frac{1}{1 + x^3} = (1 + x)(1 - x^3 + x^6 - x^9 + x^{12} - \cdots).$$

$$(1)$$

(This formula, of course, makes sense only if x has absolute value less than 1, but we can restrict our attention to those values of x if we are only interested in the coefficients of polynomials in x.) Since $x^{2n+1}(1 - x)^{2n+1}$ contains only terms of degree higher than $2n$, the required sum is the coefficient of x^{2n} in the product (1). It follows from this that

$$\binom{2n}{0} - \binom{2n-1}{1} + \binom{2n-2}{2} - \cdots + (-1)^n \binom{n}{n}$$

$$= \begin{cases} 1 \text{ for } 2n = 6k, & \text{that is, for } n = 3k, \\ 0 \text{ for } 2n = 6k + 2, & \text{that is, for } n = 3k + 1, \\ -1 \text{ for } 2n = 6k + 4, & \text{that is, for } n = 3k + 2. \end{cases}$$

57i. The expression to be evaluated equals the coefficient of x^n in the polynomial

$$(1 + x)^{2n} + 2(1 + x)^{2n-1} + 2^2(1 + x)^{2n-2} + \cdots + 2^n(1 + x)^n.$$

By the formula for the sum of a geometric progression,

$$(1 + x)^{2n} + 2(1 + x)^{2n-1} + 2^2(1 + x)^{2n-2} + \cdots + 2^n(1 + x)^n$$

$$= (1 + x)^{2n} \left[1 + \frac{2}{1 + x} + \frac{2^2}{(1 + x)^2} + \cdots + \frac{2^n}{(1 + x)^n} \right]$$

$$= (1 + x)^{2n} \frac{\dfrac{2^{n+1}}{(1 + x)^{n+1}} - 1}{\dfrac{2}{1 + x} - 1} = (1 + x)^{2n} \frac{\dfrac{2^{n+1}}{(1 + x)^n} - (1 + x)}{2 - (1 + x)}$$

$$= [2^{n+1}(1 + x)^n - (1 + x)^{2n+1}] \frac{1}{1 - x}.$$

But for $|x| < 1$,

$$\frac{1}{1 - x} = 1 + x + x^2 + x^3 + \cdots$$

Consequently, the required sum is the coefficient of x^n in the expression

$$2^{n+1}(1 + x)^n(1 + x + x^2 + \cdots) - (1 + x)^{2n+1}(1 + x + x^2 + \cdots)$$

If we multiply any polynomial $P(x) = a_0 + a_1x + a_2x^2 + \cdots + a_Nx^N$ by $1 + x + x^2 + \cdots$, then the coefficient of x^n in the resulting expression will be $a_0 + a_1 + a_2 + \cdots + a_n$, where we put $a_k = 0$ if $k > N$. In fact, the terms of degree n in the product are obtained by multiplying each term a_kx^k of $P(x)$ $(0 \leq k \leq n)$ by the term x^{n-k} in the sum $1 + x + x^2 + \cdots$. Thus, after multiplying out the product $2^{n+1}(1 + x)^n(1 + x + x^2 + \cdots)$, the coefficient of x^n will be the sum of the coefficients of the polynomial $2^{n+1}(1 + x)^n$, that is, $2^{n+1}(1 + 1)^n = 2^{2n+1}$ (compare with part a). The coefficient of x^n in the product $(1 + x)^{2n+1}(1 + x + x^2 + \cdots)$ is the sum of the coefficients of $x^0 = 1, x, x^2, \ldots, x^n$ in the polynomial $(1 + x)^{2n+1}$, that is, the sum of the first half of the coefficients of the polynomial $(1 + x)^{2n+1}$. But since $\binom{2n}{k} = \binom{2n}{2n - k}$, this sum is equal to half the sum of the coefficients of $(1 + x)^{2n+1}$; that is, it equals $\frac{1}{2} \cdot 2^{2n+1} = 2^{2n}$. It follows from this that the coefficient of x^n in the expression

$$2^{n+1}(1 + x)^n(1 + x + x^2 + \cdots) - (1 + x)^{2n+1}(1 + x + x^2 + \cdots)$$

is $2^{2n+1} - 2^{2n} = 2^{2n}$.

57j. The expression to be evaluated equals the coefficient of x^n in the following product:

$$\left\{ \binom{n}{0} + \binom{n}{1}x + \binom{n}{2}x^2 + \cdots + \binom{n}{n}x^n \right\}$$
$$\times \left\{ \binom{n}{n} + \binom{n}{n - 1}x + \binom{n}{n - 2}x^2 + \cdots + \binom{n}{0}x^n \right\}.$$

But $\binom{n}{n - k} = \binom{n}{k}$, and consequently,

$$\binom{n}{n} + \binom{n}{n - 1}x + \binom{n}{n - 2}x^2 + \cdots + \binom{n}{0}x^n$$
$$= \binom{n}{0} + \binom{n}{1}x + \binom{n}{2}x^2 + \cdots + \binom{n}{n}x^n$$
$$= (1 + x)^n.$$

We therefore have only to find the coefficient of x^n in the product $(1 + x)^n(1 + x)^n = (1 + x)^{2n}$. This coefficient is clearly $\binom{2n}{n}$.

57k. Here we have to determine the coefficient of x^n in the product

$$\left\{ \binom{n}{0} - \binom{n}{1}x + \binom{n}{2}x^2 - \cdots + (-1)^n \binom{n}{n}x^n \right\}$$

$$\cdot \left\{ \binom{n}{n} + \binom{n}{n-1}x + \binom{n}{n-2}x^2 + \cdots + \binom{n}{0}x^n \right\}.$$

This product equals $(1-x)^n(1+x)^n = (1-x^2)^n$. Therefore the coefficient of x^n is 0 when n is odd, while if $n = 2m$ is even, it is $(-1)^m \binom{2m}{m}$.

57 l. It follows from the identity

$$(1+x)^n(1+x)^m = (1+x)^{n+m}$$

that the expression equals $\binom{n+m}{k}$ (compare with the solutions to parts j and k).

58a, b. From problems 57a and 57b we have

$$\binom{n}{0} + \binom{n}{1} + \binom{n}{2} + \binom{n}{3} + \cdots + \binom{n}{n} = 2^n$$

$$\binom{n}{0} - \binom{n}{1} + \binom{n}{2} - \binom{n}{3} + \cdots + (-1)^n \binom{n}{n} = \begin{cases} 0 \text{ if } n > 0 \\ 1 \text{ if } n = 0 \end{cases}.$$

Adding these two equations and dividing the sum by 2, we obtain

$$\binom{n}{0} + \binom{n}{2} + \binom{n}{4} + \cdots = \begin{cases} 2^{n-1} \text{ if } n > 0 \\ 1 \quad \text{ if } n = 0, \end{cases}$$

which gives the answer to part a.

Subtracting the second equation from the first and dividing the difference by 2, we get

$$\binom{n}{1} + \binom{n}{3} + \binom{n}{5} + \cdots = 2^{n-1} \qquad (n \geq 1),$$

which gives the answer to part b.

58c, d, e, f. We recall that if $i = \sqrt{-1}$, then $i^2 = -1, i^3 = -i, i^4 = 1$, $i^5 = i$, and in general

$$i^k = \begin{cases} 1 \text{ if } k \text{ is of the form } 4l \\ i \text{ if } k \text{ is of the form } 4l + 1 \\ -1 \text{ if } k \text{ is of the form } 4l + 2 \\ -i \text{ if } k \text{ is of the form } 4l + 3 \end{cases}$$

Therefore, using the binomial theorem to evaluate the expressions $(1 + 1)^n$, $(1 + i)^n$, $(1 - 1)^n$, and $(1 - i)^n$, we obtain

(1) $\dbinom{n}{0} + \dbinom{n}{1} + \dbinom{n}{2} + \dbinom{n}{3} + \cdots + \dbinom{n}{n} = 2^n$

(2) $\dbinom{n}{0} + i\dbinom{n}{1} - \dbinom{n}{2} - i\dbinom{n}{3} + \cdots + i^n\dbinom{n}{n} = (1 + i)^n$

(3) $\dbinom{n}{0} - \dbinom{n}{1} + \dbinom{n}{2} - \dbinom{n}{3} + \cdots + (-1)^n\dbinom{n}{n} = \begin{cases} 0 \text{ if } n > 0 \\ 1 \text{ if } n = 0 \end{cases}$

(4) $\dbinom{n}{0} - i\dbinom{n}{1} - \dbinom{n}{2} + i\dbinom{n}{3} + \cdots + (-i)^n\dbinom{n}{n} = (1 - i)^n.$

Adding these four equations and dividing by 4, we get

$$\dbinom{n}{0} + \dbinom{n}{4} + \dbinom{n}{8} + \cdots = \begin{cases} \dfrac{2^n + (1 + i)^n + (1 - i)^n}{4} & \text{if } n > 0 \\ 1 & \text{if } n = 0 \end{cases}$$

This solves part c, but the answer for $n > 0$ can be further simplified. For let us write the complex numbers $1 + i$ and $1 - i$ in trigonometric form:

$$1 + i = \sqrt{2}\,(\cos 45° + i \sin 45°)$$
$$1 - i = \sqrt{2}\,(\cos 45° - i \sin 45°).$$

By de Moivre's theorem[1] it follows that

$$(1 + i)^n = (\sqrt{2})^n(\cos n \cdot 45° + i \sin n \cdot 45°),$$
$$(1 - i)^n = (\sqrt{2})^n(\cos n \cdot 45° - i \sin n \cdot 45°).$$

so that

$$(1 + i)^n + (1 - i)^n = 2(\sqrt{2})^n \cos n \cdot 45° = 2^{\frac{n+2}{2}} \cos n \cdot 45°.$$

Now

$$\cos n \cdot 45° = \begin{cases} 1 & \text{if } n \text{ is of the form } 8k \\ \dfrac{1}{\sqrt{2}} & \text{if } n \text{ is of the form } 8k \pm 1 \\ 0 & \text{if } n \text{ is of the form } 8k \pm 2 \\ -\dfrac{1}{\sqrt{2}} & \text{if } n \text{ is of the form } 8k \pm 3 \\ -1 & \text{if } n \text{ is of the form } 8k + 4 \end{cases}$$

[1] This is the name of formula $(\cos \alpha + i \sin \alpha)^n = \cos n\alpha + i \sin n\alpha$. To prove it one needs only to use the identity $(\cos \alpha + i \sin \alpha)(\cos \beta + i \sin \beta) = \cos (\alpha + \beta) + i \sin (\alpha + \beta)$ (which follows directly from the addition formulas for $\sin x$ and $\cos x$) and apply mathematical induction.

Therefore, if $n \geq 1$ we have

$$\binom{n}{0} + \binom{n}{4} + \binom{n}{8} + \cdots = \begin{cases} 2^{n-2} + 2^{\frac{n-2}{2}} & \text{for } n = 8k \\ 2^{n-2} + 2^{\frac{n-3}{2}} & \text{for } n = 8k \pm 1 \\ 2^{n-2} & \text{for } n = 8k \pm 2 \\ 2^{n-2} - 2^{\frac{n-3}{2}} & \text{for } n = 8k \pm 3 \\ 2^{n-2} - 2^{\frac{n-2}{2}} & \text{for } n = 8k + 4 \end{cases}$$

To solve part d we multiply equations (1), (2), (3), and (4) by 1, $-i$, -1, and i respectively, and add. This gives (noting that $n \geq 1$),

$$4\left\{\binom{n}{1} + \binom{n}{5} + \binom{n}{9} + \cdots\right\} = 2^n - i(1+i)^n + i(1-i)^n.$$

Proceeding as in part c we find that

$$\binom{n}{1} + \binom{n}{5} + \binom{n}{9} + \cdots = 2^{n-2} + 2^{\frac{n-2}{2}} \sin n \cdot 45°$$

$$= \begin{cases} 2^{n-2} & \text{for } n = 8k \quad\text{ or }\quad n = 8k + 4 \\ 2^{n-2} + 2^{\frac{n-3}{2}} & \text{for } n = 8k + 1 \quad\text{ or }\quad n = 8k + 3 \\ 2^{n-2} + 2^{\frac{n-2}{2}} & \text{for } n = 8k + 2 \\ 2^{n-2} - 2^{\frac{n-3}{2}} & \text{for } n = 8k - 1 \quad\text{ or }\quad n = 8k - 3 \\ 2^{n-2} - 2^{\frac{n-2}{2}} & \text{for } n = 8k - 2 \end{cases}$$

To solve part e, multiply equations (1), (2), (3), and (4) by 1, -1, 1, and -1 respectively, and add. This gives

$$4\left\{\binom{n}{2} + \binom{n}{6} + \binom{n}{10} + \cdots\right\} = 2^n - (1+i)^n - (1-i)^n,$$

from which we deduce as before that

$$\binom{n}{2} + \binom{n}{6} + \binom{n}{10} + \cdots = 2^{n-2} - 2^{\frac{n-2}{2}} \cos n \cdot 45°$$

$$= \begin{cases} 2^{n-2} - 2^{\frac{n-2}{2}} & \text{for } n = 8k \\ 2^{n-2} - 2^{\frac{n-3}{2}} & \text{for } n = 8k \pm 1 \\ 2^{n-2} & \text{for } n = 8k \pm 2 \\ 2^{n-2} + 2^{\frac{n-3}{2}} & \text{for } n = 8k \pm 3 \\ 2^{n-2} + 2^{\frac{n-2}{2}} & \text{for } n = 8k + 4 \end{cases}$$

Finally, to solve part f we multiply equations (1), (2), (3), and (4) by 1, i, -1, and $-i$ respectively, and add. This gives

$$4\left\{\binom{n}{3} + \binom{n}{7} + \binom{n}{11} + \cdots\right\} = 2^n + i(1 + i)^n - i(1 - i)^n.$$

Proceeding as before, we find that

$$\binom{n}{3} + \binom{n}{7} + \binom{n}{11} + \cdots = 2^{n-2} - 2^{\frac{n-2}{2}} \sin n \cdot 45°$$

$$= \begin{cases} 2^{n-2} & \text{for } n = 8k & \text{or } n = 8k + 4 \\[2mm] 2^{n-2} - 2^{\frac{n-3}{2}} & \text{for } n = 8k + 1 & \text{or } n = 8k + 3 \\[2mm] 2^{n-2} - 2^{\frac{n-2}{2}} & \text{for } n = 8k + 2 \\[2mm] 2^{n-2} + 2^{\frac{n-3}{2}} & \text{for } n = 8k - 1 & \text{or } n = 8k - 3 \\[2mm] 2^{n-2} + 2^{\frac{n-2}{2}} & \text{for } n = 8k - 2 \end{cases}$$

Remark. Part e could also be solved by subtracting c from a, and part f could be solved by subtracting d from b.

58g, h, i. These problems are solved in much the same way as the four preceding ones; only instead of the numbers 1, i, -1, $-i$, which are the four fourth roots of unity (that is, the four solutions of the equation $x^4 = 1$), we use here the three cube roots of unity (the three solutions of the equation $x^3 = 1$), which have the values

$$1, \quad \frac{-1 + i\sqrt{3}}{2}, \quad \text{and} \quad \frac{-1 - i\sqrt{3}}{2}.$$

We will use the notation $\omega = (-1 + i\sqrt{3})/2$; then $\omega^2 = (-1 - i\sqrt{3})/2$. Note further that $1 + \omega + \omega^2 = 0$, and that

$$\omega^k = \begin{cases} 1 & \text{if } k = 3l \\ \omega & \text{if } k = 3l + 1 \\ \omega^2 & \text{if } k = 3l + 2 \end{cases}$$

Applying the binomial theorem to evaluate $(1 + 1)^n$, $(1 + \omega)^n$, and $(1 + \omega^2)^n$, we obtain

$$(5) \quad \binom{n}{0} + \binom{n}{1} + \binom{n}{2} + \binom{n}{3} + \cdots + \binom{n}{n} = 2^n$$

$$(6) \quad \binom{n}{0} + \omega\binom{n}{1} + \omega^2\binom{n}{2} + \binom{n}{3} + \cdots + \omega^n\binom{n}{n} = (1 + \omega)^n$$

$$(7) \quad \binom{n}{0} + \omega^2\binom{n}{1} + \omega\binom{n}{2} + \binom{n}{3} + \cdots + \omega^{2n}\binom{n}{n} = (1 + \omega^2)^n.$$

Adding these three equations, we get

$$3\left\{\binom{n}{0} + \binom{n}{3} + \binom{n}{6} + \cdots\right\} = 2^n + (1 + \omega)^n + (1 + \omega^2)^n.$$

Let us now write the complex numbers $1 + \omega$ and $1 + \omega^2$ in trigonometric form:

$$1 + \omega = \frac{1 + i\sqrt{3}}{2} = \cos 60° + i \sin 60°$$

$$1 + \omega^2 = \frac{1 - i\sqrt{3}}{2} = \cos 60° - i \sin 60°.$$

Applying de Moivre's formula, we obtain

$$(1 + \omega)^n = \cos n \cdot 60° + i \sin n \cdot 60°$$
$$(1 + \omega^2)^n = \cos n \cdot 60° - i \sin n \cdot 60°,$$

so that

$$2^n + (1 + \omega)^n + (1 + \omega^2)^n = 2^n + 2 \cos n \cdot 60°.$$

Now

$$\cos n \cdot 60° = \begin{cases} 1 & \text{if } n \text{ is of the form } 6k \\ \tfrac{1}{2} & \text{if } n \text{ is of the form } 6k \pm 1 \\ -\tfrac{1}{2} & \text{if } n \text{ is of the form } 6k \pm 2 \\ -1 & \text{if } n \text{ is of the form } 6k + 3. \end{cases}$$

Therefore

$$\binom{n}{0} + \binom{n}{3} + \binom{n}{6} + \cdots = \tfrac{1}{3}(2^n + 2 \cos n \cdot 60°)$$

$$= \begin{cases} \tfrac{1}{3}(2^n + 2) & \text{for } n = 6k \\ \tfrac{1}{3}(2^n + 1) & \text{for } n = 6k \pm 1 \\ \tfrac{1}{3}(2^n - 1) & \text{for } n = 6k \pm 2 \\ \tfrac{1}{3}(2^n - 2) & \text{for } n = 6k + 3. \end{cases}$$

To solve part h, multiply equations (5), (6), and (7) by 1, ω^2, and ω respectively, and add. This gives

$$3\left\{\binom{n}{1} + \binom{n}{4} + \binom{n}{7} + \cdots\right\} = 2^n + \omega^2(1 + \omega)^n + \omega(1 + \omega^2)^n.$$

Since $1 + \omega + \omega^2 = 0$, we have $\omega^2 = -(1 + \omega)$, and $\omega = -(1 + \omega^2)$. Therefore

$$\omega^2(1 + \omega)^n = -(1 + \omega)^{n+1} = -\cos (n + 1)60° - i \sin (n + 1)60°$$
$$\omega(1 + \omega^2)^n = -(1 + \omega^2)^{n+1} = -\cos (n + 1)60° + i \sin (n + 1)60°.$$

This gives

$$2^n + \omega^2(1 + \omega)^n + \omega(1 + \omega^2)^n = 2^n - 2\cos(n+1)60°.$$

Therefore

$$\binom{n}{1} + \binom{n}{4} + \binom{n}{7} + \cdots = \tfrac{1}{3}\{2^n - 2\cos(n+1)60°\}$$

$$= \begin{cases} \tfrac{1}{3}(2^n - 1) & \text{for } n = 6k \quad \text{ or } n = 6k - 2 \\ \tfrac{1}{3}(2^n + 1) & \text{for } n = 6k + 1 \text{ or } n = 6k + 3 \\ \tfrac{1}{3}(2^n + 2) & \text{for } n = 6k + 2 \\ \tfrac{1}{3}(2^n - 2) & \text{for } n = 6k - 1. \end{cases}$$

To solve part i we multiply equations (5), (6), and (7) by 1, ω, and ω^2 respectively, and add. This gives

$$3\left\{\binom{n}{2} + \binom{n}{5} + \binom{n}{8} + \cdots\right\} = 2^n + \omega(1 + \omega)^n + \omega^2(1 + \omega^2)^n.$$

Proceeding as above we find easily that

$$\binom{n}{2} + \binom{n}{5} + \binom{n}{8} + \cdots = \tfrac{1}{3}\{2^n + 2\cos(n+2)60°\}$$

$$= \begin{cases} \tfrac{1}{3}(2^n - 1) & \text{for } n = 6k \quad \text{ or } n = 6k + 2 \\ \tfrac{1}{3}(2^n - 2) & \text{for } n = 6k + 1 \\ \tfrac{1}{3}(2^n + 1) & \text{for } n = 6k + 3 \text{ or } n = 6k - 1 \\ \tfrac{1}{3}(2^n + 2) & \text{for } n = 6k - 2. \end{cases}$$

This result also follows by subtracting g and h from the equation

$$\binom{n}{0} + \binom{n}{1} + \binom{n}{2} + \cdots + \binom{n}{n} = 2^n.$$

59. We will prove the theorem by mathematical induction on n. When $n = 0$ both sides of the formula reduce to 1, and therefore the theorem is true in that case. Now suppose that it is true for some integer $n \geqq 0$, that is, that

$$(a + b)^{n\,|\,h} = a^{n\,|\,h} + \binom{n}{1}a^{(n-1)\,|\,h}b^{1\,|\,h} + \binom{n}{2}a^{(n-2)\,|\,h}b^{2\,|\,h} + \cdots + b^{n\,|\,h}$$

$$\tag{1}$$

We must then show that the theorem is also true for $n + 1$. To do this, multiply both sides of equation (1) by $a + b - nh$. On the left-hand side we obtain $(a + b)^{(n+1)\,|\,h}$, as can be seen at once from the definition.

On the right-hand side we obtain a sum whose i-th term (where i runs from 0 to n) is

$$\binom{n}{i} a^{(n-i)\,|\,h} b^{i\,|\,h}(a + b - nh) = \binom{n}{i} a^{(n-i)\,|\,h} b^{i\,|\,h}(a - (n - i)h)$$
$$+ \binom{n}{i} a^{(n-i)\,|\,h} b^{i\,|\,h}(b - ih)$$
$$= \binom{n}{i} a^{(n-i+1)\,|\,h} b^{i\,|\,h} + \binom{n}{i} a^{(n-i)\,|\,h} b^{(i+1)\,|\,h}.$$

Summing over i we get

$$\binom{n}{0} a^{(n+1)\,|\,h} + \binom{n}{1} a^{n\,|\,h} b^{1\,|\,h} + \cdots + \binom{n}{n} a^{1\,|\,h} b^{n\,|\,h}$$
$$+ \binom{n}{0} a^{n\,|\,h} b^{1\,|\,h} + \cdots + \binom{n}{n-1} a^{1\,|\,h} b^{n\,|\,h} + \binom{n}{n} b^{(n+1)\,|\,h}.$$

Using the relation $\binom{n}{i-1} + \binom{n}{i} = \binom{n+1}{i}$, this becomes

$$a^{(n+1)\,|\,h} + \binom{n+1}{1} a^{n\,|\,h} b^{1\,|\,h} + \binom{n+1}{2} a^{(n-1)\,|\,h} b^{2\,|\,h} + \cdots + b^{(n+1)\,|\,h}.$$
$$(2)$$

Thus we have shown that $(a + b)^{(n+1)\,|\,h}$ is equal to the expression (2), which is precisely the statement of the theorem for $n + 1$. This completes the induction.

60a. Using the fact that

$$\binom{n}{i} = \frac{n(n-1)\cdots(n-i+1)}{i!} = \frac{n^{i\,|\,1}}{i!},$$

we obtain

$$\binom{n}{i}\binom{m}{k-i} = \frac{n^{i\,|\,1}}{i!} \frac{m^{(k-i)\,|\,1}}{(k-i)!}$$
$$= \frac{1}{k!} \frac{k!}{i!\,(k-i)!} n^{i\,|\,1} m^{(k-i)\,|\,1}$$
$$= \frac{1}{k!}\binom{k}{i} n^{i\,|\,1} m^{(k-i)\,|\,1}.$$

Consequently the sum to be evaluated equals

$$\frac{1}{k!}\left\{\binom{k}{0} m^{k\,|\,1} + \binom{k}{1} m^{(k-1)\,|\,1} n^{1\,|\,1} + \binom{k}{2} m^{(k-2)\,|\,1} n^{2\,|\,1}\right.$$
$$\left. + \cdots + \binom{k}{k} n^{k\,|\,1}\right\} = \frac{(m+n)^{k\,|\,1}}{k!} = \binom{m+n}{k}.$$

This solution was obtained before in a different way (see problem 57 l).

60b. The i-th term of the sum to be evaluated (where i runs from 0 to k) can be rewritten as follows:

$$(-1)^i \binom{m+i}{i}\binom{n}{k-i} = (-1)^i \frac{(m+i)^{i|1}}{i!}\frac{n^{(k-i)|1}}{(k-i)!}$$

$$= \frac{1}{k!}\binom{n}{i}(-1)^i(m+i)^{i|1}n^{(k-i)|1}.$$

Now

$$(-1)^i(m+i)^{i|1} = (-1)^i(m+i)(m+i-1)\cdots(m+1)$$
$$= (-m-1)(-m-2)\cdots(-m-i+1)(-m-i)$$
$$= (-m-1)^{i|1}.$$

Consequently,

$$(-1)^i \binom{m+i}{i}\binom{n}{k-i} = \frac{1}{k!}\binom{k}{i}(-m-1)^{i|1}n^{(k-i)|1}.$$

Therefore our sum equals

$$\frac{1}{k!}\left\{\binom{k}{0}n^{k|1} + \binom{k}{1}n^{(k-1)|1}(-m-1)^{1|1}\right.$$

$$+ \binom{k}{2}n^{(k-2)|1}(-m-1)^{2|1} + \cdots + \left.\binom{k}{k}(-m-1)^{k|1}\right\}$$

$$= \frac{(n-m-1)^{k|1}}{k!} = \frac{(n-m-1)(n-m-2)\cdots(n-m-k)}{k!}.$$

If $n-m-1 \geq k$, this equals $\binom{n-m-1}{k}$.

Note that since $\binom{m+i}{i} = \binom{m+i}{m}$ we can rewrite this identity in the form

$$\binom{m}{m}\binom{n}{k} - \binom{m+1}{m}\binom{n}{k-1} + \binom{m+2}{m}\binom{n}{k-2}$$

$$- \cdots + (-1)^k\binom{m+k}{m}\binom{n}{0}$$

$$= \frac{(n-m-1)(n-m-2)\cdots(n-m-k)}{k!}.$$

61a. Consider the shortest paths which lead from the intersection (0,0) to the intersection $(n-m+1, m)$ in fig. 58. The number of such paths is

$\binom{n+1}{m}$. Let us now divide these paths into the following pairwise disjoint classes. The first class shall consist of those paths which start along the "horizontal" street; there will be as many such paths as there are paths joining the intersection (1,0) to the intersection $(n-m+1, m)$, that is, $\binom{n}{m}$. The second class shall consist of those paths which start by going one block up the "vertical" street to the intersection (0,1) and then turn right; the number of such paths will be the number of paths joining the intersection (1,1) to the intersection $(n-m+1, m)$, that is, $\binom{n-1}{m-1}$.

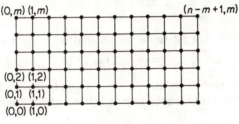

Fig. 58

The third class shall consist of the paths which start by going two blocks up the vertical street to the intersection (0,2) and then turn right; the number of such paths is equal to the number of paths joining the points (1,2) and $(n-m+1, m)$, that is, $\binom{n-2}{m-2}$. The fourth class shall consist of all the paths which start by going three blocks up the "vertical" street to the intersection (0,3) and then turn right (a total of $\binom{n-3}{m-3}$ paths), etc. The last class consists of the path which goes m blocks up the vertical street and then turns right (there is $\binom{n-m}{0} = 1$ such path). Since each of the $\binom{n+1}{0}$ shortest paths connecting (0,0) to $(n-m+1, m)$ belongs to exactly one of the classes considered above, the relation to be proved follows immediately.

Note that this relation can also be derived easily from the result of problem 57g. For since $\binom{n}{m} = \binom{n}{n-m}$, the relation to be proved can be rewritten in the form

$$\binom{n}{n-m} + \binom{n-1}{n-m} + \binom{n-2}{n-m} + \cdots + \binom{n-m}{n-m} = \binom{n+1}{m}.$$

Setting $n = m + k$ and reversing the order of the terms in the left-hand side, we obtain:

$$\binom{k}{k} + \binom{k+1}{k} + \binom{k+2}{k} + \cdots + \binom{k+m}{k} = \binom{m+k+1}{m}$$
$$= \binom{m+k+1}{k+1}.$$

But this is a special case of problem 57g.

It is also possible to derive the required relation directly from the binomial theorem: this amounts to determining the coefficients of x^m in the polynomial

$$(1+x)^n + x(1+x)^{n-1} + x^2(1+x)^{n-2} + \cdots + x^m(1+x)^{n-m},$$

which is the sum of $m + 1$ terms of a geometric progression with ratio $x/(1+x)$.

61b. Consider all shortest routes which lead from the point (0,0) to the point $(n - m + k + 1, m)$. The number of such paths is $\binom{n+k+1}{m}$. We will divide these paths into pairwise disjoint classes. To do this, we draw a vertical line between the k-th and $(k + 1)$st vertical streets (fig. 59).

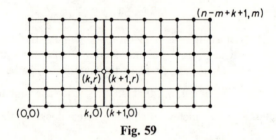

Fig. 59

Each of our paths meets this line in a single point. The number of paths which meet this line at a point on the r-th horizontal street is the product of the number of paths joining the point (0,0) to the point (k,r) and the number of paths joining the point $(k + 1, r)$ to the point $(n - m + k + 1, m)$. This number therefore equals $\binom{k+r}{r}\binom{n-r}{m-r}$. Setting $r = 0, 1, 2, \ldots, m$ and adding the expressions obtained, we find that

$$\binom{k}{0}\binom{n}{m} + \binom{k+1}{1}\binom{n-1}{m-1} + \binom{k+2}{2}\binom{n-2}{m-2} + \cdots$$
$$+ \binom{k+m}{m}\binom{n-m}{0} = \binom{n+k+1}{m},$$

as was to be shown.

In the special case $k = 0$, we obtain

$$\binom{n}{m} + \binom{n-1}{m-1} + \cdots + \binom{n-m}{0} = \binom{n+1}{m},$$

which is the result of part a. For $k = 1$, we obtain the following relation:

$$\binom{n}{m} + 2\binom{n-1}{m-1} + \cdots + (m+1)\binom{n-m}{0} = \binom{n+2}{m}.$$

61c. Consider again our geometric diagram. $\binom{m}{k}\binom{n}{0}$ is the product of the number of shortest paths from the point $(0,0)$ to the point $(m-k, k)$ and the number of shortest paths from the point $(m-k, k)$ to the point $(n+m-k, k)$; this product is the number of shortest paths from the point $(0,0)$ to the point $(n+m-k, k)$ which pass through the point

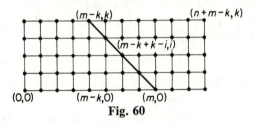

Fig. 60

$(m-k, k)$ (fig. 60). Similarly, $\binom{m}{k-1}\binom{n}{1}$ is the number of shortest paths from the point $(0,0)$ to the point $(n+m-k, k)$ which pass through the point $(m-k+1, k-1)$; $\binom{m}{k-2}\binom{n}{2}$ is the number of shortest paths from the point $(0,0)$ to the point $(n+m-k, k)$ which pass through the point $(m-k+2, k-2)$, etc. and $\binom{m}{0}\binom{n}{k}$ is the number of shortest paths from the point $(0,0)$ to the point $(n+m-k, k)$ which pass through the point $(m-k+k, 0) = (m,0)$. But since each of the shortest paths from the point $(0,0)$ to the point $(n+m-k, k)$ must intersect the line drawn in fig. 60 at exactly one point, the sum involved in problem 61c is the total number of shortest paths joining the points $(0,0)$ and $(n+m-k, k)$; hence it equals $\binom{n+m}{k}$.

For other applications of the method used here, see the solutions to problems 83a–c below.

62. We will at once solve the more general problem b. Select any path in our network of roads which connects the point A to one of the intersections of the thousandth row; we will determine how many people take this path. The path in question consists of 1000 separate "blocks"; moreover, at the end of each block the people who have been walking along that block split into two groups: half of them continue on the path in question and half of them turn off it. It is clear that of the $2^{1000}/2 = 2^{999}$ people who start out along the path, only $2^0 = 1$ keeps on it until the end. Thus, exactly one of the 2^{1000} persons takes each of the possible paths in the network. It follows from this that the total number of paths is 2^{1000} (which can also be verified without difficulty by a direct computation). The problem thus amounts to that of determining how many different paths lead to each intersection of the thousandth row. But the network of roads represented in fig. 7 (see p. 19) is exactly the same as the network of roads represented in fig. 6; only in fig. 6 the "streets" go in the horizontal and vertical directions and in fig. 7 they go in the directions L and R. For a person to arrive at the k-th intersection (counting from left to right and calling the leftmost intersection the 0-th of the 1000th row, he would have to go k blocks in the direction R (and the remaining $1000 - k$ blocks in the direction L). Consequently, the total number of paths which lead to the k-th intersection is $\binom{1000}{k}$. Thus $\binom{1000}{k} = 1000!/[k!(1000 - k)!]$ people arrive at the k-th intersection, where $0 \leq k \leq 1000$.

From this result it follows in particular that the numbers of people who arrive at the three leftmost crossings B_1, B_2, and B_3 are respectively $\binom{1000}{0} = 1$, $\binom{1000}{1} = 1000$, and $\binom{1000}{2} = (1000 \cdot 999)/2 = 499{,}500$. This same result could also have been obtained by direct computation.

63. The three relations to be proved can easily be derived from the fact that the number $B_n{}^k$ is the coefficient of x^k in the polynomial $(1 + x + x^2)^n$:

$$(1 + x + x^2)^n = B_n{}^0 + B_n{}^1 x + B_n{}^2 x^2 + \cdots + B_n{}^{2n} x^{2n}. \tag{1}$$

Let us first prove this formula.

For $n = 0$,

$$(1 + x + x^2)^0 = 1;$$

that is, the formula holds (since $B_n{}^0 = 1$). Suppose now that this formula has already been established for the exponent n; let us show that in this case it will also hold for the exponent $n + 1$. For this purpose, let us multiply both sides of equation (1) by $(1 + x + x^2)$. On the left side we get $(1 + x + x^2)^{n+1}$, and on the right side we get

$$(B_n{}^0 + B_n{}^1 x + B_n{}^2 x^2 + B_n{}^3 x^3 + \cdots + B_n{}^{2n} x^{2n})(1 + x + x^2)$$
$$= B_n{}^0 + (B_n{}^0 + B_n{}^1)x + (B_n{}^0 + B_n{}^1 + B_n{}^2)x^2$$
$$+ (B_n{}^1 + B_n{}^2 + B_n{}^3)x^3 + \cdots + (B_n{}^{2n-1} + B_n{}^{2n})x^{2n+1} + B_n{}^{2n}x^{2n+2}.$$

By virtue of the rule for forming the numbers B_{n+1}^k, the latter expression equals

$$B_{n+1}^0 + B_{n+1}^1 + B_{n+1}^2 x^2 + B_{n+1}^3 x^3 + \cdots + B_{n+1}^{2n+2} x^{2n+2}.$$

We have thus demonstrated that if our formula holds for the exponent n, it also holds for the exponent $n + 1$. Since the formula holds for $n = 0$, it must therefore hold for all n.

Note further that it follows from the symmetry of the triangular array of numbers that $B_n^k = B_n^{2n-k}$ for arbitrary n and $k \leqq 2n$.

Now the required relations can be proved immediately in exactly the same way that the relations of problems 57a, b, and j were proved.

a. $B_n^0 + B_n^1 + B_n^2 + \cdots + B_n^{2n} = (1 + 1 + 1)^n = 3^n.$

b. $B_n^0 - B_n^1 + B_n^2 - \cdots + B_n^{2n} = (1 - 1 + 1)^n = 1.$

c. The sum to be evaluated is the coefficient of x^{2n} in the polynomial

$$(B_n^0 + B_n^1 x + B_n^2 x^2 + \cdots + B_n^{2n} x^{2n})$$
$$\times (B_n^{2n} + B_n^{2n-1} x + B_n^{2n-2} x + \cdots + B_n^0 x^{2n}).$$

But since $B_n^{2n-k} = B_n^k$, this polynomial equals

$$(B_n^0 + B_n^1 x + B_n^2 x^2 + \cdots + B_n^{2n} x^{2n})^2 = (1 + x + x^2)^{2n},$$

from which the required relation follows immediately.

VI. PROBLEMS ON COMPUTING PROBABILITIES

64. We are as likely to encounter any one of the 10,000 bicycles as any other; hence there are a total of 10,000 equally likely possible outcomes to this experiment. It remains only to compute the number of favorable outcomes, that is, the number of bicycles whose license numbers do not contain the digit 8. This number can be determined with the aid of the following method. Let us prefix to each number of less than four digits enough zeros to make four figures (for example, instead of 26 we will write 0026), and instead of 10,000 let us write 0000. In doing this, we have not changed the number of license numbers which contain no 8's among their digits. But the numbers which we how have (0000, 0001, 0002, . . . , 9999) constitute all possible four-digit numbers. If one of these numbers does not contain the digit 8, then the possibilities for its first digit are

0, 1, 2, . . . , 7, 9; these are likewise all possibilities for the second, third, and fourth digits. Combining each of the nine possible values for the first digit with each of the nine possible values for the second digit, each of the nine possible values for the third, and each of the nine possible values for the fourth, we obtain a total of 9^4 four-digit numbers which do not contain the digit 8 (compare with the solution to problem 11).

Thus, of the 10,000 equally likely outcomes (one corresponding to each of the 10,000 bicycles which one might first encounter), 9^4 are favorable, from which it follows that the required probability is $9^4/10,000 = (0.9)^4 = 0.6561$.

65a. The experiment under consideration consists of drawing four cards at random from a given set of six cards. There are six possible outcomes for the drawing of the first card; corresponding to each of these six possibilities there are five possible outcomes for the drawing of the second card (any of the five remaining cards can be drawn); corresponding to each of these outcomes there are four possible outcomes to the drawing of the third card; and corresponding to each of the latter outcomes there are three possible outcomes to the drawing of the fourth card. Combining each of the six possibilities for the first card with each of the corresponding five possibilities for the second card, we obtain $6 \cdot 5 = 30$ possible outcomes for the drawing of the first two cards; similarly there are $30 \cdot 4 = 120$ possible outcomes for the drawing of the first three cards and $120 \cdot 3 = 360$ possible outcomes for the drawing of all four cards. Thus the experiment can lead to any of 360 different outcomes; these outcomes are all equally likely.

Exactly one of these outcomes is favorable, namely, that in which the D is picked on the first draw, the E on the second draw, the A on the third draw, and the F on the fourth draw. Consequently, the required probability is $1/360 \approx 0.003$.

65b. Here again the experiment consists of drawing four cards in succession from a set of six cards; consequently there are 360 possible outcomes to the experiment. But here the computation of the number of favorable outcomes is more complicated than in the case of part a since some letters are repeated in the set of cards. The favorable outcomes are those in which one of the three D cards is picked on the first draw, one of the two O cards on the second draw, one of the two remaining D cards on the third draw, and the one remaining O card on the fourth draw. Combining each of the three possibilities for the first card, each of the two possibilities for the second card, each of the two possibilities for the third card, and the single possibility for the fourth card, we obtain a total of $3 \cdot 2 \cdot 2 \cdot 1 = 12$ different favorable outcomes. Consequently, the probability to be computed is $12/360 = 1/30 \approx 0.033$.

66. Let the numbers drawn be (in the order of drawing) a, b, c, d, e. The number obtained is then

$$N = 10{,}000a + 1{,}000b + 100c + 10d + e.$$

As in the preceding problem we see that the number of different possible outcomes of the experiment is $10 \cdot 9 \cdot 8 \cdot 7 \cdot 6 = 30{,}240$. Let us now determine the number of favorable outcomes.

In order that N be divisible by $495 = 5 \cdot 9 \cdot 11$, it is necessary and sufficient that it be divisible by 5, by 9, and by 11. Since

$$10{,}000a + 1{,}000b + 100c + 10d$$

is always a multiple of 5, N is divisible by 5 if and only if $e = 0$ or $e = 5$.
Since

$$N = (9999a + 999b + 99c + 9d) + (a + b + c + d + e),$$

and since the first parenthesis is always a multiple of 9, N is divisible by 9 if and only if $a + b + c + d + e$ is.[1]

Finally, since

$$N = (99{,}999a + 1001b + 99c + 11d) + (a - b + c - d + e),$$

and since the first parenthesis is a multiple of 11, N is divisible by 11 if and only if $a - b + c - d + e$ is.

Now

$$a + b + c + d + e \geqq 0 + 1 + 2 + 3 + 4 = 10,$$

and

$$a + b + c + d + e \leqq 9 + 8 + 7 + 6 + 5 = 35.$$

Therefore $a + b + c + d + e$ can be divisible by 9 only if it equals 18 or 27. Let us consider these cases separately.

1. Let $a + b + c + d + e = 18$. Then $a - b + c - d + e$ is even (since it is equal to $18 - 2b - 2d$) and $|a - b + c - d + e| < 18$. Consequently

$$a - b + c - d + e = 0,$$

since 0 is the only even multiple of 11 in the permitted range. Therefore

$$a + c + e = b + d = 9.$$

If $e = 0$, then $a + c = b + d = 9$. Hence (a,c) and (b,d) must be chosen from the eight pairs $(1,8), (2,7), (3,6), \ldots, (8,1)$. Once (a,c) is chosen, there are only six possibilities left for (b,d), since it cannot be equal

---

[1] The same reasoning shows that any positive integer N is divisible by 9 if and only if the sum of its digits (in the decimal scale) is divisible by 9. This is the rule of "casting out 9's". Similarly it can be shown that N is divisible by 11 if and only if the alternating sum of its digits (i.e., the first digit minus the second plus the third, etc.) is divisible by 11. See, e.g., Hardy and Wright, *An Introduction to the Theory of Numbers*, New York, Oxford, 1960, p. 114.

to (a,c) or (c,a). Therefore we obtain a total of $8 \cdot 6 = 48$ solutions when $e = 0$.

If $e = 5$, then $a + c = 4$, $b + d = 9$. Therefore (a,c) must be one of the pairs $(0,4)$, $(4,0)$, $(1,3)$, or $(3,1)$. In the first two cases, (b,d) can be any one of the eight pairs $(0,9)$, $(1,8)$, $(2,7)$, $(3,6)$, $(6,3)$, $(7,2)$, $(8,1)$, or $(9,0)$. But in the last two cases (b,c) must be one of the four pairs $(0,9)$, $(2,7)$, $(7,2)$, or $(9,0)$. Therefore we obtain a total of $2 \cdot 8 + 2 \cdot 4 = 24$ solutions when $e = 5$.

Thus the total number of favorable outcomes in case 1 is $48 + 24 = 72$.

2. Let $a + b + c + d + e = 27$. In this case $a - b + c - d + e$ is odd (since it equals $27 - 2b - 2d$). Moreover,

$$27 > a - b + c - d + e > 27 - 2 \cdot 9 - 2 \cdot 8 = -7.$$

Hence $a - b + c - d + e = 11$, since there are no other odd multiples of 11 in the permitted range.

It follows that $b + d = 8$ and $a + c + e = 19$. Here $e = 0$ is impossible, since $9 + 8 < 19$, and so we must have $e = 5$. This gives $a + c = 14$, so that (a,c) must be one of the pairs $(6,8)$ or $(8,6)$. This eliminates the values 5, 6, and 8 for b and d, and therefore (b,d) must be either $(1,7)$ or $(7,1)$. Hence we obtain $2 \cdot 2 = 4$ favorable outcomes in this case.

Combining cases 1 and 2 we get a total of $72 + 4 = 76$ favorable outcomes. The desired probability is therefore $76/30{,}240 = 19/7560 \approx 0.0025$.

67. The experiment in question consists of selecting the forgotten digit at random; if this yields a wrong number, then another digit is chosen at random. Thus the possible outcomes fall into two classes: those in which two numbers are chosen and those in which only one is chosen. (The second class contains only one outcome: that in which the right number is chosen the first time.) However, we have no reason to suppose that the first kind of outcome and the second kind are equally probable; in fact they are not. To obtain a system of equally probable outcomes it is convenient to assume that two numbers are picked *in all cases*, that is, that even when the boy gets the right number the first time, he makes a second call. (This, of course, would not be done in practice; but by considering this fictitious experiment we can simplify the solution. The probability of getting the right number is the same in both cases—making a second call has no bearing on the outcome of the first call.)

The revised experiment, which consists merely of making two different random choices of the forgotten digit, has $10 \cdot 9 = 90$ equally probable possible outcomes (any of 10 numbers can be chosen the first time, and the second time any number can be drawn except the one drawn the first time). The favorable outcomes are obviously the nine in which the right number comes up the first time and some other number the second time, and the nine outcomes in which one of the nine wrong numbers comes up the first time and the right number comes up the second time. Thus the total number of favorable outcomes is $9 + 9 = 18$. Consequently, the required probability is $18/90 = 1/5 = 0.2$.

68a. Here the experiment consists of having 12 people tell in what months their birthdays occur. The birthday of the first person can fall in any of the 12 months, and the second person's birthday can likewise occur in any of the 12 months. Combining each of the 12 possibilities for the first person with each of the 12 possibilities for the second person, we obtain $12 \cdot 12 = 12^2$ equally likely possible outcomes for the first two people. Similarly we will have 12^3 equally likely possible outcomes for the first three people, and so forth; for all 12 people we obtain 12^{12} possibilities.

Let us now compute how many of these outcomes are favorable, that is, such that the birthdays all occur in different months. In a favorable outcome, the first person's birthday can occur in any month, but the second person's birthday must then occur in one of the 11 other months, the third person's birthday in one of the 10 remaining months, etc., up to the last person, whose birthday must occur in the one month which remains. Combining each of the 12 possibilities for the first person with each of the corresponding 11 possibilities for the second person, each of the $12 \cdot 11$ possibilities thus obtained for the first two people with each of the corresponding 10 possibilities for the third person, etc., we obtain a total of $12 \cdot 11 \cdot 10 \cdots 1 = 12!$ favorable outcomes. Consequently, the required probability is $12!/12^{12} \approx 0.000054$.

68b. The total number of possible outcomes is obtained here as in part a and equals 12^6. Let us now compute the number of favorable outcomes. The number of outcomes in which the birthdays of all six people occur in two given months (say, January and April) is 2^6; this number is obtained by exactly the same reasoning as we used in obtaining the number 12^6 for the total number of possible outcomes. Of these 2^6 outcomes, we must discard the two in which the birthdays of all six people occur either all in the one month (January) or all in the other month (April). These two outcomes are not favorable, since in them the six birthdays are not distributed over two months, but lie in just one month. Thus, the number of favorable outcomes in which all the birthdays lie in two specific months equals $2^6 - 2$. Since one can choose two months out of the 12 in $\binom{12}{2} = 66$ different ways, the total number of favorable outcomes is

$66 \cdot (2^6 - 2) = 66 \cdot 62$. It follows from this that the required probability is $(66 \cdot 62)/12^6 \approx 0.00137$.

69. The first passenger can choose any of the three cars. The second passenger can then likewise choose any of the three cars, so that the total number of ways two passengers can choose which cars to sit in is $3 \cdot 3 = 3^2$. Similarly, the number of ways nine passengers can choose which cars to sit in is 3^9. There are therefore 3^9 equally likely possible outcomes. Let us now determine the number of favorable outcomes for parts a, b, and c.

69a. The three passengers who sit in the first car can be chosen from the nine passengers in $\binom{9}{3} = \dfrac{9 \cdot 8 \cdot 7}{1 \cdot 2 \cdot 3}$ different ways. The number of ways of arranging the remaining six passengers in the other two cars is 2^6 (by the same argument we used in showing that the nine passengers could be arranged in three cars in 3^9 ways). Combining each of the $\binom{9}{3}$ ways of choosing which three passengers will sit in the first car with each of the 2^6 ways of seating the other six passengers in the remaining two cars, we get a total of $\binom{9}{3} \cdot 2^6 = \dfrac{9 \cdot 8 \cdot 7 \cdot 2^6}{1 \cdot 2 \cdot 3}$ favorable outcomes. It follows from this that the required probability is

$$\frac{9 \cdot 8 \cdot 7 \cdot 2^6}{1 \cdot 2 \cdot 3 \cdot 3^9} = \frac{1792}{6561} \approx 0.273.$$

69b. The number of ways of choosing which three passengers will sit in the first car is the number of combinations of nine objects three at a time, which equals $\binom{9}{3} = \dfrac{9 \cdot 8 \cdot 7}{1 \cdot 2 \cdot 3}$. (See solution to part a.) Any three of the six remaining passengers can take seats in the second car; one can choose these three passengers in $\binom{6}{3} = \dfrac{6 \cdot 5 \cdot 4}{1 \cdot 2 \cdot 3}$ ways. Combining each of the $\binom{9}{3}$ ways of choosing which three passengers will sit in the first car with each of the $\binom{6}{3}$ ways of choosing which three of the other passengers will sit in the second car, we obtain a total of $\binom{9}{3}\binom{6}{3} = \dfrac{9 \cdot 8 \cdot 7 \cdot 6 \cdot 4}{(1 \cdot 2 \cdot 3)^2} = \dfrac{9!}{(3!)^3}$ favorable outcomes.

Thus the required probability is

$$\frac{9!}{(3!)^3 3^9} = \frac{560}{6561} \approx 0.085.$$

69c. The car in which two passengers sit can be chosen in three ways and the passengers who sit there can be chosen in $\binom{9}{2}$ ways. Then the car in which three passengers sit can be chosen in two ways, and the passengers who sit there can be chosen in $\binom{7}{3}$ ways. The other passengers must then sit in the remaining car. Hence there are $3 \cdot 2 \binom{9}{2}\binom{7}{3} = 7560$ favorable outcomes, and the required probability is $7560/3^9 \approx 0.384$. (This probability is $4\frac{1}{2}$ times as great as that of the arrangement in part b.)

70. The experiment considered in this problem consists of dividing the ten cards into two sets A and B of five cards each. The total number of ways of choosing the set A is $\binom{10}{5} = 252$, and so there are 252 equally likely possible outcomes. We must now determine the number of favorable outcomes for parts a, b, and c.

70a. The number of outcomes in which the 9 and 10 are both in A is equal to the number of ways of choosing the three remaining cards of A from among the 1, 2, . . . , 8. This can be done in $\binom{8}{3} = 56$ ways.

By the same reasoning there are 56 outcomes in which the 9 and 10 are both in B. Therefore the required probability is $(56 + 56)/252 = 4/9$.

This problem can also be solved without computing the total number of outcomes. To do this, note that whichever hand the 10 is in, the 9 can either be one of the four remaining cards in that hand or one of the five cards in the other hand, thus giving four favorable and five unfavorable outcomes.

70b. The number of outcomes in which the 8, 9, and 10 are all in hand A is equal to the number of ways of choosing the two remaining cards of A from among the 1, 2, . . . , 7. This can be done in $\binom{7}{2} = 21$ ways. Similarly there are 21 outcomes with the 8, 9, and 10 in hand B, so that the required probability is

$$\frac{21 + 21}{252} = \frac{1}{6}.$$

70c. There are $\binom{4}{2} = 6$ ways of deciding which two of the top four cards are to be in hand A. Once this is determined the three remaining cards of A must be chosen from among the 1, 2, . . . , 6. This can be done in $\binom{6}{3} = 20$ ways. Therefore there are $6 \cdot 20 = 120$ favorable outcomes, and the desired probability is $120/252 = 10/21$.

71. We will compute the two probabilities in the problem and then compare them.

1. The total number of possible outcomes in a set of four games is $2^4 = 16$, since each game has two possible outcomes. These are all equally likely, since A and B are equally strong. The favorable outcomes are those in which B wins only one game out of the four; this can happen in four ways, and therefore the first probability is $4/16 = 1/4$.

2. In a set of eight games there are $2^8 = 256$ equally likely possible outcomes. The favorable outcomes are those in which B wins only three of the games. The three he wins can be chosen in $\binom{8}{3} = 56$ ways. Thus the second probability is $56/256 = 7/32$.

Since $7/32 < 1/4$, A is more likely to win three games out of four than to win five games out of eight.[2]

72a. We consider here an experiment which consists of drawing k balls from a box containing $n + m$ balls. Since k balls can be selected from the $n + m$ in $\binom{n+m}{k}$ different ways, there are $\binom{n+m}{k}$ equally likely possible outcomes to our experiment. The favorable ones are those in which exactly r of the k balls drawn are white, and consequently, $k - r$ of them black. It is clear that this is possible only when $r \leq k, r \leq n$, and $k - r \leq m$; when these conditions do not all hold, there are no favorable outcomes at all and the desired probability is zero. Assuming that the indicated inequalities are satisfied, we obtain all the favorable outcomes by combining each of the $\binom{n}{r}$ ways of drawing r balls from the n white balls with the $\binom{m}{k-r}$ ways of drawing $k - r$ balls from the m black balls. Thus the number of favorable outcomes is $\binom{n}{r}\binom{m}{k-r}$, and the required probability is $\binom{n}{r}\binom{m}{k-r} \bigg/ \binom{n+m}{k}$.

72b. Let $k \leq n$ and $k \leq m$. When k balls are drawn from a box containing n white balls and m black ones, the number of white balls drawn can be $0, 1, 2, \ldots$, or k. According to part a, the probability that this number will be 0 is $\binom{n}{0}\binom{m}{k} \bigg/ \binom{n+m}{k}$, the probability that it will be 1 is

---

[2] This answer seems at first paradoxical, since $3/4 > 5/8$, which would seem to imply that it is harder to win three out of four games from an equally strong opponent than to win five games out of eight. The point is that one must distinguish between the probability of winning *exactly* five games out of eight, which is what the problem deals with, and the probability of winning *at least* five games out of eight. It is in fact easier to win at least five games out of eight from an equally strong opponent than to win at least three games out of four.

$\binom{n}{1}\binom{m}{k-1}\Big/\binom{n+m}{k}$. The probability is $\binom{n}{2}\binom{m}{k-2}\Big/\binom{n+m}{k}$ that it will be 2, ..., and the probability is $\binom{n}{k}\binom{m}{0}\Big/\binom{n+m}{k}$ that it will be k. But the sum of these probabilities is 1, since exactly one of the above events must occur. Thus

$$\frac{\binom{n}{0}\binom{m}{k}}{\binom{n+m}{k}} + \frac{\binom{n}{1}\binom{m}{k-1}}{\binom{n+m}{k}} + \frac{\binom{n}{2}\binom{m}{k-2}}{\binom{n+m}{k}} + \cdots + \frac{\binom{n}{k}\binom{m}{0}}{\binom{n+m}{k}} = 1,$$

and consequently

$$\binom{n}{0}\binom{m}{k} + \binom{n}{1}\binom{m}{k-1} + \binom{n}{2}\binom{m}{k-2} + \cdots + \binom{n}{k}\binom{m}{0} = \binom{n+m}{k}.$$

Remark. The method used in problem 72b is closely related to the geometric method discussed in problem 61. In both cases we divide some set of objects (the shortest paths in the geometric method and possible outcomes to the experiment in the solution of problem 72b), the number of whose elements is to be computed, into mutually disjoint subsets, and then equate the number of elements in the entire set to the sum of the numbers of elements in the subsets. But the method of problem 72b is more general and can be applied to derive many relations which are essentially unobtainable by the geometric method of problem 61 (see, for example, problem 73).

73a. Our problem amounts to determining the probability that when the man tries to take an $(n+1)$st match from box A, exactly $n-k$ matches have been taken from the box B. In this form, the problem is not changed if we increase the number of matches in the two boxes; in fact we can suppose that there are an infinite number of matches in each box. The advantage of this is that now we can consider the experiment of choosing $n + (n-k) = 2n-k$ matches, selecting the box from which each is taken at random (as the $2n-k+1$st match is assumed to come from A, its selection is not part of the experiment). There are 2^{2n-k} equally likely outcomes for this experiment. The favorable outcomes are those in which n matches are chosen from A and the remaining $n-k$ from B. This can be done in $\binom{2n-k}{n}$ ways, and so the desired probability is

$$\frac{1}{2^{2n-k}}\binom{2n-k}{n}.$$

73b. At the moment when the man in part a discovers that he has picked an empty box, the number of matches in the other box can be anything from 0 to n. The sum of the probabilities of these events is 1, and therefore

$$\frac{1}{2^{2n}}\binom{2n}{n} + \frac{1}{2^{2n-1}}\binom{2n-1}{n} + \frac{1}{2^{2n-2}}\binom{2n-2}{n} + \cdots + \frac{1}{2^{n}}\binom{n}{n} = 1.$$

Multiplying by 2^{2n}, we get

$$\binom{2n}{n} + 2\binom{2n-1}{n} + 2^2\binom{2n-2}{n} + \cdots + 2^n\binom{n}{n} = 2^{2n}.$$

74. *First solution.* Let us define a sequence $\{a_1, a_2, \ldots, a_{50}\}$ by putting $a_n = 1$ if A hits a duck on his n-th shot and $a_n = 0$ otherwise. Similarly we define $\{b_1, b_2, \ldots, b_{51}\}$ by putting $b_n = 1$ if B hits a duck on his n-th shot and 0 otherwise. The combined sequence $S = \{a_1, a_2, \ldots, a_{50}, b_1, b_2, \ldots, b_{51}\}$ then completely describes the outcome of the hunt. Since the probability of hitting a duck on any given shot is 1/2 for both A and B, there are 2^{101} equally likely possible outcomes $\{a_1, \ldots, a_{50}, b_1, \ldots, b_{51}\}$. For any given sequence, A's total number of hits is $a_1 + a_2 + \cdots + a_{50}$ and B's total is $b_1 + b_2 + \cdots + b_{51}$. Therefore an outcome is favorable if and only if

$$a_1 + a_2 + \cdots + a_{50} < b_1 + b_2 + \cdots + b_{51}.$$

We have to determine the number of sequences with this property.

To every sequence $S = \{a_1, a_2, \ldots, a_{50}, b_1, b_2, \ldots, b_{51}\}$ we can associate the sequence $S' = \{1 - a_1, 1 - a_2, \ldots, 1 - a_{50}, 1 - b_1, 1 - b_2, \ldots, 1 - b_{51}\}$ obtained by interchanging the zeros and ones in S. We will now show that S is favorable if and only if S' is unfavorable. If S is favorable, then

$$a_1 + a_2 + \cdots + a_{50} < b_1 + b_2 + \cdots + b_{51},$$

and so

$$-a_1 - a_2 - \cdots - a_{50} > -b_1 - b_2 > \cdots > -b_{51},$$

since when an inequality is multiplied by -1, its sense is reversed. Adding 51 to both sides, we obtain

$$(1 - a_1) + (1 - a_2) + \cdots + (1 - a_{50}) + 1$$
$$> (1 - b_1) + (1 - b_2) + \cdots + (1 - b_{51}).$$

But if x and y are integers, then $x + 1 > y$ is equivalent to $x \geq y$. Therefore

$$(1 - a_1) + (1 - a_2) + \cdots + (1 - a_{50})$$
$$\geq (1 - b_1) + (1 - b_2) + \cdots + (1 - b_{51}).$$

This shows that S' is unfavorable. The reasoning is reversible, and so we have established a one-to-one correspondence between favorable and unfavorable outcomes. There are therefore 2^{100} of each, and the desired probability is 1/2.

Note that we do not actually need to know the total number of outcomes to solve this problem. For once we show that F, the number of favorable outcomes, is equal to U, the number of unfavorable outcomes, the required probability is $F/(F + U) = F/(F + F) = 1/2$.

Second solution. Let us imagine that the hunt takes place in two stages: in the first stage A and B each shoot at 50 ducks, and in the second stage B shoots at one duck. There are then two ways in which B can hit more ducks than A:

(1) by hitting more ducks than A in the first stage (the outcome of the second stage is then irrelevant);

(2) by hitting the same number of ducks as A in the first stage and then hitting the duck in the second stage.

Let p be the probability that (1) occurs, i.e., that B hits more ducks than A in the first stage. By symmetry p is also the probability that A hits more ducks than B in the first stage (indeed, both A and B have the same accuracy and the same number of shots in the first stage). Therefore the probability of a tie at the end of the first stage is $1 - 2p$. This implies that the probability of (2) is $(1 - 2p)\frac{1}{2}$, for the two stages are independent of each other (see p. 22). Since (1) and (2) are mutually exclusive, the probability that one or the other occurs is $p + (1 - 2p)\frac{1}{2} = 1/2$.

Remark. The same reasoning applies to a hunt in which A shoots at n ducks and B shoots at $n + 1$ ducks. The probability that B bags more ducks than A is always $1/2$.

75a. The experiment we are discussing in this problem deals with two hunters simultaneously shooting at a fox. The frequency of hits and misses by each of the hunters of course does not depend on the result of the simultaneous shot by the other: if they shoot simultaneously at a fox many times, the first hunter will hit the fox with an average of one shot out of three, and the second hunter will hit the fox just as often when the first hunter is successful as when he is not. In computing the probability, we can suppose that instead of shooting, each hunter independently draws a slip of paper from a hat containing three slips, one marked "Hit" and the other two marked "Miss."

Call the "Hit" slip H and the two "Miss" slips M_1 and M_2; combining each of the three possible outcomes for the first hunter's drawing with each of the three possible outcomes for the second hunter's drawing, we obtain the following nine possibilities for our experiment:

$$(M_1\ M_1)\quad (M_1\ M_2)\quad (M_1\ H)$$
$$(M_2\ M_1)\quad (M_2\ M_2)\quad (M_2\ H)$$
$$(H\ M_1)\quad (H\ M_2)\quad (H\ H).$$

Of these nine possibilities, the favorable ones are those in which at least one of the hunters draws the "Hit" slip, that is, the five outcomes in the bottom row and rightmost column of the above table. Thus the probability that at least one of the hunters hits the fox is 5/9.

75b. Combining each of the nine possible outcomes of the preceding problem with each of the three possible outcomes of a third hunter's drawing one of the slips, we obtain 27 equally likely possible outcomes. Of these 27 possible outcomes, one of the first two hunters draws the Hit slip in $3 \cdot 5 = 15$ cases (remember that in the previous part the Hit slip was drawn in five cases out of nine); in the remaining 12 cases the first two people draw Miss slips in each case and the third person draws the Hit slip in one-third of the cases (that is, in four cases). Thus there are a total of $15 + 4 = 19$ favorable outcomes. Hence the probability that at least one of the three hunters will hit the fox is 19/27.

75c. In the case of n hunters firing simultaneously, we obtain as in parts a and b 3^n equally likely possible outcomes. A direct calculation of the number F of favorable outcomes can be made (using the principle of inclusion and exclusion), but is somewhat complicated. It is simpler to compute the number U of unfavorable outcomes, that is, those in which all n hunters miss. We then have $F = 3^n - U$. To find U, it is again convenient to replace the actual shooting by drawing slips as in parts a and b.

To create an unfavorable outcome each hunter must draw either M_1 or M_2. Thus there are two possibilities for each of the n hunters, and so $U = 2^n$. Therefore $F = 3^n - 2^n$, and the required probability is

$$(3^n - 2^n)/3^n = 1 - (2/3)^n.$$

Remark: This problem can also be solved by working directly with probabilities instead of computing F or U. For if p is the probability that at least one hunter scores a hit, then $q = 1 - p$ is the probability that they all miss. The probability that any one hunter misses is 2/3; therefore by the remarks on independent events made in the introduction to this section, $q = (2/3)^n$, and $p = 1 - (2/3)^n$. This solution is substantially the same as that given above.

76. Let us first compute the probability of the second or third shot hitting the fox. The second time the hunter would be shooting from a distance of 150 yards and the third time from a distance of 200 yards. Since by hypothesis the probability of a hit is inversely proportional to the square of the distance, and is 1/2 when the distance is 100 yards, the probability of a hit on the second shot is $(1/2)(100/150)^2 = 2/9$, and on the third shot it is $(1/2)(100/200)^2 = 1/8$.

From these calculations we see that the probability of missing with the first shot is $1 - 1/2 = 1/2$, that of missing with the second shot is

$1 - 2/9 = 7/9$, and that of missing with the third shot is $1 - 1/8 = 7/8$. Therefore the probability of missing with all three shots is $1/2 \cdot 7/9 \cdot 7/8 = 49/144$, and so the solution to the problem is $1 - 49/144 = 95/144 \approx 0.66$.

77. We will solve the problem in the formulation given in the remark. Let us suppose, however, that instead of being given a blank sheet of paper, A is given a slip already marked with a plus. He can then either leave it alone or change it to a minus, and the probability that he leaves it alone is known to be $1/3$.

The experiment under discussion consists of passing the paper from A to B to C to D and observing the final result. Let X be the event that A left the plus sign and let Y be the event that the final result was a plus sign. We have to find $Pr\{X \mid Y\}$, the conditional probability of X given Y. This quantity is equal to $Pr\{X \cap Y\}/Pr\{Y\}$ (see the introduction to this section). To calculate $Pr\{Y\}$ observe that in order for the final sign to be plus it must have changed an even number of times (since it started out as a plus). Therefore it changed either 0, 2, or 4 times. The probability that it changed 0 times is $(1/3)^4 = 1/81$, and the probability that it changed 4 times is $(2/3)^4 = 16/81$. To calculate the probability that it changed twice, note that there are $\binom{4}{2} = 6$ ways to pick the two people who changed it. For each choice of these people, the probability that they changed it and the other two left it alone is $(2/3)^2(1/3)^2 = 4/81$. Thus the probability of exactly two changes is $6 \cdot 4/81 = 24/81$. Therefore $Pr\{Y\} = 1/81 + 24/81 + 16/81 = 41/81$.

For the event $X \cap Y$ to occur there must have been an even number of changes, but A did not make one of them. Hence there were either 0 or 2 changes. As before, the probability of 0 changes is $(1/3)^4 = 1/81$, but now the probability of 2 changes is only $\binom{3}{2}(2/3)^2(1/3)^2 = 12/81$, since the people making the changes can only be chosen in $\binom{3}{2} = 3$ ways. Therefore

$$Pr\{X \cap Y\} = 1/81 + 12/81 = 13/81,$$

and so

$$Pr\{X \mid Y\} = (13/81)/(41/81) = 13/41.$$

Remark. The above problem can be generalized as follows. Let n people $A_1, \ldots, A_n$ be given. A slip marked with a plus sign is given to A_1, who passes it to A_2, who passes it to A_3, etc.; finally A_n passes it to a judge. At the i-th stage A_i has the option of changing the sign before passing it on; assume that each A_i exercises this option with probability p, where $0 < p < 1$.

Now suppose that the judge observes a plus sign at the end of the process. What is the probability that A_1 left the sign unchanged? (We leave it to the reader to formulate this as a problem of n liars.)

For convenience of notation, put $q = 1 - p$; thus q is the probability that A_i does not change the sign.

Let X be the event that A_1 does not change the sign, and let Y be the event that the final sign is a plus. We must calculate $Pr\{X \mid Y\} = Pr\{X \cap Y\}/Pr\{Y\}$.

To find $Pr\{Y\}$, note that in order for the final sign to be plus, it must have changed an even number $2k$ of times, where $0 \leqq 2k \leqq n$. Let us calculate the probability that the sign is changed exactly $2k$ times. There are $\binom{n}{2k}$ ways to choose which people make the changes. Suppose these people have been chosen; let us say they are $A_1, \ldots, A_{2k}$ for definiteness. The probability that they will all change the sign, but that the remaining people $A_{2k+1}, \ldots, A_n$ will leave it alone, is $\overbrace{p \cdot p \cdots p}^{2k} \overbrace{q \cdot q \cdots q}^{n-2k} = p^{2k}q^{n-2k}$. The same reasoning applies to whatever group of $2k$ people make the changes, so the desired probability is $\binom{n}{2k} p^{2k}q^{n-2k}$. Therefore

$$Pr\{Y\} = \binom{n}{0}q^n + \binom{n}{2}p^2q^{n-2} + \binom{n}{4}p^4q^{n-4} + \cdots,$$

where the series breaks off as soon as $2k$ becomes greater than n. To evaluate this in closed form, apply the binomial theorem to $(q + p)^n$ and $(q - p)^n$. We get

$$(q + p)^n = \binom{n}{0}q^n + \binom{n}{1}q^{n-1}p + \binom{n}{2}q^{n-2}p^2 + \binom{n}{3}q^{n-3}p^3 + \cdots + \binom{n}{n}p^n$$

$$(q - p)^n = \binom{n}{0}q^n - \binom{n}{1}q^{n-1}p + \binom{n}{2}q^{n-2}p^2 \\ - \binom{n}{3}q^{n-3}p^3 + \cdots + (-1)^n\binom{n}{n}p^n.$$

Adding these two equations and then dividing by 2, we see that

$$\frac{(q + p)^n + (q - p)^n}{2} = \binom{n}{0}q^n + \binom{n}{2}q^{n-2}p^2 + \binom{n}{4}q^{n-4}p^4 + \cdots \\ = Pr\{Y\}.$$

Since $q + p = 1$, this can be simplified to give $Pr\{Y\} = [1 + (1 - 2p)^n]/2$.

Next we must find $Pr\{X \cap Y\}$. In order for $X \cap Y$ to occur, an even number of people must change the sign, but A_1 must not be one of them. So if exactly $2k$ people change the sign, they can be chosen in only $\binom{n-1}{2k}$ ways.

Once the $2k$ people are chosen, the probability that they will change the sign while the other $n - 2k$ people will leave it alone is still $p^{2k}q^{n-2k}$. Therefore

$$Pr\{X \cap Y\} = \binom{n-1}{0}q^n + \binom{n-1}{2}p^2q^{n-2} + \binom{n-1}{4}p^4q^{n-4} + \cdots \\ = q\left\{\binom{n-1}{0}q^{n-1} + \binom{n-1}{2}p^2q^{n-3} + \binom{n-1}{4}p^4q^{n-5} + \cdots\right\}.$$

The expression in the braces can be evaluated as above, and we get

$$Pr\{X \cap Y\} = q \frac{1 + (1 - 2p)^{n-1}}{2}.$$

Finally $\qquad Pr\{X \mid Y\} = \dfrac{Pr\{X \cap Y\}}{Pr\{Y\}} = (1 - p) \dfrac{1 + (1 - 2p)^{n-1}}{1 + (1 - 2p)^n}.$

The reasoning used in this solution can be applied to a great many problems in probability theory, in particular in the theory of *Markov chains*.[3]

78a. The upper ends of the six blades of grass can be joined in pairs in $5 \cdot 3 \cdot 1 = 15$ different ways (the first end can be tied to any of the other five upper ends; then the first loose upper end can be tied to any of the other three loose ends; then the two remaining loose ends must be tied together). There are likewise 15 different ways of joining the lower ends. Since the way the lower ends are joined is independent of the way the upper ends are joined there are a total of $15 \cdot 15 = 225$ equally likely possible outcomes to the experiment.

Let us now compute the number of favorable outcomes. Let the upper ends be connected in any of the 15 possible ways; let, say, the end of the first blade be tied to the end of the second blade, the third to the fourth, and the fifth to the sixth. In order that a ring be obtained, it is necessary that the lower end of the first blade be tied to the lower end of the third, fourth, fifth, or sixth blade; we thus have four possibilities for the lower end of the first blade. Further, if the lower end of the first blade is joined to that of the third blade, then the lower end of the second blade will have to be joined to that of either the fifth or the sixth blade; here we have only two possibilities. After this is done we are left with only two loose ends, which must be joined to each other. Combining all possibilities, we see that for each of the 15 ways of joining the upper ends there are exactly $4 \cdot 2 = 8$ ways of joining the lower ends which lead to favorable outcomes to the experiment. It follows from this that the total number of favorable outcomes is $15 \cdot 8 = 120$.

Thus the probability to be computed is $(15 \cdot 8)/(15 \cdot 15) = 8/15 \approx 0.53$.

78b. By the same argument as in the solution to part a, we find that when there are $2n$ blades of grass, the total number of possible outcomes to the experiment is $[(2n - 1)(2n - 3)(2n - 5) \cdots 1]^2$, and the number of favorable outcomes is

$$[(2n - 1)(2n - 3)(2n - 5) \cdots 1][(2n - 2)(2n - 4) \cdots 2].$$

[3] See J. G. Kemeny and J. L. Snell, *Finite Markov Chains*, Princeton, 1960.

Thus the required probability is

$$p_n = \frac{(2n-2)(2n-4)\cdots 2}{(2n-1)(2n-3)\cdots 3\cdot 1} = \frac{2n-2}{2n-1}\cdot\frac{2n-4}{2n-3}\cdots\frac{2}{3}.$$

Remark. The apparently simple answer to problem 78b becomes very inconvenient for large values of n, since in that case a large number of fractions must be multiplied. However, by using Stirling's formula[4]

$$n! \sim n^n e^{-n}\sqrt{2\pi n}$$

one can obtain a simple approximation to this probability which is useful for large n. For noting that $(2n-2)(2n-4)\cdots 2 = 2^{n-1}(n-1)\cdot(n-2)\cdots 1 = 2^{n-1}(n-1)!$, and multiplying the numerator and denominator of the answer to problem 78b by this expression, we find that

$$p_n = \frac{[2^{n-1}(n-1)!]^2}{(2n-1)(2n-2)!} \sim \frac{[2^{n-1}\sqrt{2\pi(n-1)}(n-1)^{n-1}e^{-(n-1)}]^2}{(2n-1)\sqrt{2\pi\cdot 2(n-1)}\,[2(n-1)^{2(n-1)}e^{-2(n-1)}]}$$

$$= \frac{\sqrt{\pi(n-1)}}{2n-1} \sim \frac{\sqrt{\pi}}{2\sqrt{n}},$$

since for large n

$$\frac{\sqrt{n-1}}{2n-1} = \frac{\sqrt{n}\sqrt{1-\dfrac{1}{n}}}{2n\left(1-\dfrac{1}{2n}\right)} \sim \frac{\sqrt{n}}{2n} = \frac{1}{2\sqrt{n}}.$$

Here π as usual denotes the ratio $3.14\cdots$ of the circumference of a circle to its diameter. The relative error in the approximation

$$p_n \sim \frac{\sqrt{\pi}}{2\sqrt{n}}$$

decreases as n increases.

79a. Let us compute the total number of (equally likely) possible outcomes to the experiment considered in the hypothesis of the problem. The first person can draw any of $\binom{2n}{2} = \dfrac{2n(2n-1)}{2}$ pairs of balls. Then the second person can draw any of the $\binom{2n-2}{2} = (2n-2)(2n-3)/2$ pairs which can be formed from the remaining $2n-2$ balls. The third can draw any of $\binom{2n-4}{2} = (2n-4)(2n-5)/2$ pairs, etc., up to the next-to-last person, who can draw any of $\binom{4}{2} = 6$ pairs; the last person has no choice but to draw the two remaining balls. Combining each of the

[4] See, for example, R. Courant, *Differential and Integral Calculus*, Interscience, New York, 1937, p. 361–364.

$\binom{2n}{2}$ possible outcomes to the drawing of the first pair of balls with each of the $\binom{2n-2}{2}$ corresponding possible outcomes for the drawing of the second pair, then combining each of the $\binom{2n}{2}\binom{2n-2}{2}$ possibilities thus obtained for the first two pairs of balls with the $\binom{2n-4}{2}$ possible outcomes for the drawing of the third pair of balls, etc., we obtain a total of

$$\binom{2n}{2}\binom{2n-2}{2}\cdots\binom{4}{2}\binom{2}{2}$$

$$=\frac{2n(2n-1)}{2}\cdot\frac{(2n-2)(2n-3)}{2}\cdots\frac{4\cdot3}{2}\cdot\frac{2\cdot1}{2}=\frac{(2n)!}{2^n}$$

equally likely possible outcomes to the experiment. It now remains only to compute how many of these outcomes are favorable.

The favorable outcomes are those in which each person draws one white ball and one black one. The first person can draw any of the n white balls and any of the n black balls, that is, he can draw any of n^2 pairs which consist of one white ball and one black one. Then the second person can pick any of the $(n-1)^2$ remaining such pairs (after the first person has drawn one white ball and one black one, there are $n-1$ white balls and $n-1$ black ones left), then the third person can choose any of $(n-2)^2$ such pairs, ..., the next-to-last any of $2^2=4$ pairs, and the last person must pick the one remaining pair. Combining all these possibilities, we obtain a total of

$$n^2(n-1)^2(n-2)^2\cdots2^2\cdot1^2=(n!)^2$$

favorable outcomes. Consequently the required probability is

$$p_n=\frac{(n!)^2}{\dfrac{(2n)!}{2^n}}=\frac{2^n(n!)^2}{(2n)!}.$$

Remark. The seemingly simple answer obtained here turns out to be very inconvenient if one has to compute the probability for large values of n (say, for n equal to 8 or more).

As in the remark to problem 78 we can use Stirling's formula

$$n!\sim\sqrt{2\pi n}\,n^n e^{-n}$$

to estimate p_n. We get

$$p_n=\frac{2^n(n!)^2}{(2n)!}\sim\frac{2^n(2\pi n)n^{2n}e^{-2n}}{\sqrt{2\pi\cdot2n}(2n)^{2n}e^{-2n}}=\frac{\sqrt{\pi n}}{2^n}.$$

79b. There are here a total of $(2n)!/2^n$ equally likely possible outcomes (the same as in part a; see the solution to that problem). Thus we have only to find the number of favorable outcomes.

First of all, it is quite clear that for *odd n* the number of favorable outcomes (and consequently, also the required probability p_n) equals zero: if the total number of white balls is odd, at least one of the people must draw one white ball and one black ball. Hence we have only to consider the case of *even n = 2k*; the total number of possible outcomes is then equal to $(4k)!/2^{2k}$.

Let us now compute the number of favorable outcomes in which k specific people of the $2k$ participants each draw a pair of white balls (and consequently, the other k each draw a pair of black balls). The k people can each draw a pair of white balls from the $2k$ white balls in the jar in $(2k)!/2^k$ different ways (this number is obtained by replacing n by k in the expression for the total number of possible outcomes). The remaining k people can likewise each pick a pair of black balls in $(2k)!/2^k$ different ways. Combining these possibilities, we conclude that the number of outcomes in which the given k people each draw a pair of white balls is $[(2k)!]^2/2^{2k}$. But the k people who pick the white balls can be chosen from the total of $2k$ people in $\binom{2k}{2} = (2k)!/(k!)^2$ different ways. The total number of favorable outcomes is therefore

$$\frac{(2k)!^2}{2^{2k}} \cdot \frac{(2k)!}{(k!)^2} = \frac{(2k)!^3}{2^{2k}(k!)^2}.$$

Consequently, the required probability is

$$p_{2k} = \frac{(2k)!^3}{2^{2k}(k!)^2} \bigg/ \frac{(4k)!}{2^{2k}} = \frac{(2k)!^3}{(4k)!(k!)^2}.$$

Remark. As in the case of part a, for large values of k it is convenient to approximate the answer obtained by making use of Stirling's formula:

$$p_{2k} = \frac{[(2k)!]^3}{(4k)!\,(k!)^2} \sim \frac{(2\pi \cdot 2k)^{3/2}(2k)^{6k}e^{-6k}}{\sqrt{2\pi \cdot 4k}(4k)^{4k}e^{-4k}(2\pi k)k^{2k}e^{-2k}}$$

$$= \frac{(2\pi k)^{3/2} \cdot 2^{3/2} \cdot 2^{6k}k^{6k}e^{-6k}}{(2\pi k)^{3/2} \cdot 2 \cdot 4^{4k}k^{6k}e^{-6k}} = \frac{\sqrt{2}}{2^{2k}}.$$

80a. *First solution.* The experiment under discussion in this problem consists of writing the m addresses on the m envelopes. Any of the m addresses can be written on the first envelope; then any of the remaining $m - 1$ addresses can be written on the second envelope; then any of the remaining $m - 2$ addresses on the third envelope, etc. Therefore the experiment has a total of $m(m - 1)(m - 2) \cdots 1 = m!$ equally likely possible outcomes. We now have to compute the number of favorable outcomes, i.e. outcomes where at least one envelope is correctly addressed.

We will do this by applying the principle of inclusion and exclusion (see problem 12). Let A_1 be the set of all outcomes in which the first letter is correctly addressed, A_2 the set of outcomes where the second letter is correctly addressed, etc. Then $A_1 \cup A_2 \cup \cdots \cup A_m$ is the set of favorable outcomes, so our problem is to compute $\#(A_1 \cup A_2 \cup \cdots \cup A_m)$. To do this we will need to know the quantities $\#(A_i)$, $\#(A_i \cap A_j)$, $\#(A_i \cap A_j \cap A_k)$, etc. Now $\#(A_i) = (m-1)!$, since when the i-th envelope is correctly addressed, there are $(m-1)!$ ways to address the remaining $m-1$ envelopes. Similarly $\#(A_i \cap A_j) = (m-2)!$, since after the i-th and j-th envelopes are correctly addressed, the remaining $m-2$ envelopes can be addressed in $(m-2)!$ ways. Likewise $\#(A_i \cap A_j \cap A_k) = (m-3)!$, etc. In the expression on the right-hand side of the principle of inclusion and exclusion there are m terms of the form $\#(A_i)$, $\binom{m}{2}$ terms of the form $-\#(A_i \cap A_j)$, $\binom{m}{3}$ terms of the form $\#(A_i \cap A_j \cap A_k)$, etc. Therefore we have

$$\#(A_1 \cup A_2 \cup \cdots \cup A_m) = m(m-1)! - \binom{m}{2}(m-2)! + \binom{m}{3}(m-3)!$$
$$- \cdots + (-1)^{m-1}\binom{m}{m}!$$

This expression can be simplified by noting that $\binom{m}{r}(m-r)! = m!/r!$. We thus obtain

$$\#(A_1 \cup A_2 \cup \cdots \cup A_m) = m! - \frac{m!}{2!} + \frac{m!}{3!} - \cdots + (-1)^{m-1}\frac{m!}{m!}$$
$$= m!\left(1 - \frac{1}{2!} + \frac{1}{3!} - \cdots + \frac{(-1)^{m-1}}{m!}\right).$$

The required probability is obtained by dividing this number by the total number of outcomes, which we saw was $m!$ Thus the solution is $1 - 1/2! + 1/3! - \cdots + (-1)^{m-1}/m!$.

Second solution. As in the problem with the fox (problem 75c), the required probability can also be found if we compute not the total number of favorable outcomes, but the total number of *unfavorable* outcomes, that is, the outcomes in which none of the envelopes is addressed correctly. Denote the number of such outcomes by A_n. Number the envelopes in any way, using the numbers $1, 2, \ldots, n$; we will call the correct address for the k-th envelope the k-th address.

In the case of an unfavorable outcome, the possible addresses for the first envelope are the 2nd, 3rd, 4th, $\ldots$, and the n-th addresses. Consider the unfavorable outcomes in which the 2nd address is written on the first envelope. Then either the 1st address or one of the 3rd through n-th addresses can be written on the 2nd envelope. Let us consider these two cases separately.

If the 1st address is written on the 2nd envelope, then in order that the outcome be unfavorable, it is necessary and sufficient that none of the remaining $n - 2$ envelopes (the 3rd through n-th) be addressed correctly. The number of such outcomes is equal to the number of unfavorable ways of addressing $n - 2$ envelopes, that is A_{n-2}.

Now consider the outcomes in which an address other than the first is written on the second envelope. The number of such outcomes equals the number of ways in which the 1st, 3rd, 4th, . . . , and n-th addresses can be assigned to the 2nd, 3rd, 4th, . . . , and n-th envelopes in such a way that the 2nd envelope is not given the first address, the 3rd envelope is not given the 3rd address, the 4th envelope is not given the 4th address, . . . , and the n-th envelope is not given the n-th address. This number equals the number of unfavorable outcomes for the case of $n - 1$ envelopes, that is, A_{n-1} (the fact that it is here the 1st address and not the 2nd which must not be written on the 2nd envelope is completely immaterial).

Thus the total number of unfavorable outcomes in which the 2nd address is written on the 1st envelope is $A_{n-1} + A_{n-2}$. We obtain exactly the same expression for the numbers of unfavorable outcomes in which the first envelope bears the 3rd address, or the 4th adrdess, etc. Since in the unfavorable cases any of a total of $n - 1$ different addresses can be written on the first envelope, we obtain the formula

$$A_n = (n - 1)(A_{n-1} + A_{n-2}). \tag{1}$$

Consider now the probability p_n that none of the n envelopes is addressed correctly. Since in our case there are a total of $n!$ equally likely possible outcomes (see the beginning of the first solution) and the number of outcomes in which none of the envelopes is addressed correctly is A_n, we have

$$p_n = \frac{A_n}{n!}.$$

The formula (1) now gives

$$\frac{A_n}{n!} = (n - 1)\left[\frac{A_{n-1}}{n!} + \frac{A_{n-2}}{n!}\right]$$

$$= (n - 1)\left[\frac{1}{n}\frac{A_{n-1}}{(n - 1)!} + \frac{1}{n(n - 1)}\frac{A_{n-2}}{(n - 2)!}\right],$$

that is,

$$p_n = \left(1 - \frac{1}{n}\right)p_{n-1} + \frac{1}{n}p_{n-2}; \quad p_n = p_{n-1} - \frac{1}{n}(p_{n-1} - p_{n-2}).$$

For $n = 1$ there is only one possible outcome, a correct address, so that $A_1 = 0$ and $p_1 = 0$; for $n = 2$ there are two equally likely outcomes, one of them favorable and one unfavorable, so that $p_2 = 1/2$. Using the formula obtained, we can compute successively

$$p_3 = p_2 - \frac{1}{3}(p_2 - p_1) = \frac{1}{2} - \frac{1}{3} \cdot \frac{1}{2} = \frac{1}{2} - \frac{1}{2 \cdot 3},$$

$$p_4 = p_3 - \frac{1}{4}(p_3 - p_2) = \frac{1}{2} - \frac{1}{2 \cdot 3} - \frac{1}{4}\left(-\frac{1}{2 \cdot 3}\right)$$

$$= \frac{1}{2} - \frac{1}{2 \cdot 3} + \frac{1}{2 \cdot 3 \cdot 4},$$

$$p_5 = p_4 - \frac{1}{5}(p_4 - p_3) = \frac{1}{2} - \frac{1}{2 \cdot 3} + \frac{1}{2 \cdot 3 \cdot 4} - \frac{1}{5} \cdot \frac{1}{2 \cdot 3 \cdot 4}$$

$$= \frac{1}{2} - \frac{1}{2 \cdot 3} + \frac{1}{2 \cdot 3 \cdot 4} - \frac{1}{2 \cdot 3 \cdot 4 \cdot 5},$$

$$\cdots\cdots\cdots\cdots\cdots\cdots\cdots\cdots\cdots$$

$$p_n = p_{n-1} - \frac{1}{n}(p_{n-1} - p_{n-2})$$

$$= \frac{1}{2} - \frac{1}{2 \cdot 3} + \cdots + \frac{(-1)^{n-1}}{2 \cdot 3 \cdot 4 \cdots (n-1)} - \frac{1}{n} \frac{(-1)^{n-1}}{2 \cdot 3 \cdot 4 \cdots (n-1)}$$

$$= \frac{1}{2!} - \frac{1}{3!} + \frac{1}{4!} - \cdots + \frac{(-1)^{n-1}}{(n-1)!} + \frac{(-1)^n}{n!}.$$

Since the total number of *favorable* outcomes is $n! - A_n$, the probability sought in the problem (namely, that of at least one of the envelopes being addressed correctly) is

$$\frac{n! - A_n}{n!} = 1 - \frac{1}{2!} + \frac{1}{3!} - \frac{1}{4!} + \cdots - \frac{(-1)^n}{n!}.$$

80b. For large n, the sum

$$\frac{1}{2!} - \frac{1}{3!} + \cdots + (-1)^n \frac{1}{n!}$$

differs by less than $1/(n + 1)!$ from the *infinite* series

$$\frac{1}{2!} - \frac{1}{3!} + \frac{1}{4!} - \frac{1}{5!} + \cdots,$$

since

$$0 < \left(\frac{1}{(n+1)!} - \frac{1}{(n+2)!}\right) + \left(\frac{1}{(n+3)!} - \frac{1}{(n+4)!}\right) + \cdots$$

$$= \frac{1}{(n+1)!} - \left(\frac{1}{(n+2)!} - \frac{1}{(n+3)!}\right) - \cdots < \frac{1}{(n+1)!}.$$

The sum of this infinite series is $1/e$, where $e = 2.71828182 \cdots$ is the limit of the expression $(1 + 1/n)^n$ as $n \to \infty$ (e is the base of the system of natural logarithms).[5] Thus the probability found in this problem is close to $1 - 1/e \approx 0.63212056$ (that is, somewhat smaller than 2/3). Note that even for $n = 10$ our probability differs from $1 - 1/e$ only in the eighth and subsequent decimal places, since $1/11! \approx 0.00000002$.

81a. The experiment under discussion in this problem consists of each of the p passengers choosing at random (independently of the others) one of the m carriages of the train. One passenger has m possibilities for the choice of his carriage, two passengers have m^2 possibilities, ..., and p passengers have m^p possibilities. Thus there are a total of m^p equally likely possible outcomes to the experiment. We will now compute the number of *unfavorable* outcomes, i.e. outcomes in which at least one carriage is empty. Let A_i be the set of outcomes in which the i-th carriage is empty; then $A_1 \cup A_2 \cup \cdots \cup A_m$ is the set of unfavorable outcomes. We can compute $\#(A_1 \cup A_2 \cup \cdots \cup A_m)$ by the principle of inclusion and exclusion provided we know $\#(A_i)$, $\#(A_i \cap A_j)$, $\#(A_i \cap A_j \cap A_k)$, etc. Now $\#(A_i) = (m - 1)^p$, since when the i-th carriage is required to be empty, the p passengers can be put into the remaining $m - 1$ carriages in $(m - 1)^p$ ways. Similarly $\#(A_i \cap A_j) = (m - 2)^p$, for when the i-th and j-th carriages are empty, there are $m - 2$ carriages remaining in which to put the p passengers. By the same reasoning $\#(A_i \cap A_j \cap A_k) = (m - 3)^p$ etc. The principle of inclusion and exclusion therefore gives

$$\#(A_1 \cup A_2 \cup \cdots \cup A_m)$$
$$= m(m - 1)^p - \binom{m}{2}(m - 2)^p + \binom{m}{3}(m - 3)^p - \cdots + (-1)^{m-1}\binom{m}{m}0^p.$$

Subtracting this from m^p, the total number of outcomes, we see that the number of favorable outcomes is

$$m^p - \#(A_1 \cup \cdots \cup A_m)$$
$$= m^p - \binom{m}{1}(m - 1)^p + \binom{m}{2}(m - 2)^p - \cdots + (-1)^{m-1}\binom{m}{m - 1}1^p.$$

Dividing by m^p we obtain the desired probability, namely

$$\frac{m^p - \binom{m}{1}(m - 1)^p + \binom{m}{2}(m - 2)^p - \cdots + (-1)^{m-1}\binom{m}{m - 1}1^p}{m^p}$$
$$= \binom{m}{0}1^p - \binom{m}{1}\left(1 - \frac{1}{m}\right)^p + \binom{m}{2}\left(1 - \frac{2}{m}\right)^p$$
$$- \cdots + (-1)^{m-1}\binom{m}{m - 1}\left(1 - \frac{m - 1}{m}\right)^p.$$

[5] See for example R. Courant, *op. cit.*, p. 326.

81b. *First solution.*   The experiment is the same as in part a and therefore has m^p equally likely possible outcomes. But now a favorable outcome is one in which exactly r carriages are occupied. For a given set of r carriages we know from part a that the p passengers can be put into them so that none is empty in

$$\binom{r}{0}r^p - \binom{r}{1}(r-1)^p + \binom{r}{2}(r-2)^p - \cdots + (-1)^{r-1}\binom{r}{r-1}1^p$$

ways. Since r carriages can be chosen from among the m in $\binom{m}{r}$ ways, the total number of favorable outcomes is

$$\binom{m}{r}\left[\binom{r}{0}r^p - \binom{r}{1}(r-1)^p + \cdots + (-1)^{r-1}\binom{r}{r-1}1^p\right].$$

The required probability is therefore

$$\frac{\binom{m}{r}\left[\binom{r}{0}r^p - \binom{r}{1}(r-1)^p + \cdots + (-1)^{r-1}\binom{r}{r-1}1^p\right]}{m^p}.$$

Second solution.   Part b can also be solved independently of part a. Let us denote by $f(p,r)$ the number of ways in which p passengers can be arranged in m carriages so that exactly r carriages are occupied (we consider the number m to be fixed). Now consider $f(p+1, r)$. From each of the $f(p,r)$ arrangements of p passengers in which r carriages are occupied, we can obtain r such arrangements of $p+1$ passengers, since the $(p+1)$st passenger can sit in any of the r carriages already occupied. Also from each of the $f(p, r-1)$ arrangements of p passengers in which $r-1$ carriages are occupied, we can obtain $m-r+1$ arrangements of $p+1$ passengers for which r carriages are occupied. This is because the $(p+1)$st passenger can be seated in any of the $m-(r-1) = m-r+1$ hitherto unoccupied carriages. This accounts for all arrangements of $p+1$ passengers, and so $f(p+1, r) = rf(p,r) + (m-r+1)f(p, r-1)$.

To eliminate m from this equation we divide both sides by

$$m(m-1)(m-2)\cdots(m-r+1),$$

getting

$$\frac{f(p+1, r)}{m(m-1)\cdots(m-r+1)} = r\frac{f(p,r)}{m(m-1)\cdots(m-r+1)} \qquad (1)$$
$$+ \frac{f(p, r-1)}{m(m-1)\cdots(m-r+2)}.$$

Next we introduce the notation

$$\frac{f(p,r)}{m(m-1)\cdots(m-r+1)} = \left[\begin{matrix}p\\r\end{matrix}\right].$$

With the aid of this symbol, equation (1) can be written in the form

$$\begin{bmatrix} p + 1 \\ r \end{bmatrix} = r \begin{bmatrix} p \\ r \end{bmatrix} + \begin{bmatrix} p \\ r - 1 \end{bmatrix}. \qquad (2)$$

We have

$$f(1,r) = \begin{cases} m & \text{when} \quad r = 1 \\ 0 & \text{when} \quad r > 1, \end{cases}$$

because if there is only one passenger, he cannot occupy *more* than one car, and there are m ways in which he can occupy *exactly* one car. Hence

$$\begin{bmatrix} 1 \\ r \end{bmatrix} = \frac{f(1,r)}{m} = \begin{cases} 1 & \text{when} \quad r = 1 \\ 0 & \text{when} \quad r > 1 \end{cases}.$$

Equation (2) resembles the formula

$$\binom{p + 1}{r} = \binom{p}{r} + \binom{p}{r - 1}$$

for the binomial coefficients. Just as this formula can be used to construct Pascal's triangle, equation (2) can be used to construct a table of the numbers $\begin{bmatrix} p \\ r \end{bmatrix}$. The above calculation of $\begin{bmatrix} 1 \\ r \end{bmatrix}$ is needed to start the process. We obtain the array

$$
\begin{array}{cccccc}
1 \\
1 & 1 \\
1 & 3 & 1 \\
1 & 7 & 6 & 1 \\
1 & 15 & 25 & 10 & 1 \\
1 & 31 & 90 & 65 & 15 & 1 \\
\cdot & \cdot & \cdot & \cdot & \cdot & \cdot \\
\cdot & \cdot & \cdot & \cdot & \cdot & \cdot \\
\cdot & \cdot & \cdot & \cdot & \cdot & \cdot \\
\end{array},
$$

where $\begin{bmatrix} p \\ r \end{bmatrix}$ is the entry in the p-th row and r-th column. Each entry is obtained by adding the entry to the northwest and r times the entry to the north, where r is the column number. Thus, for example, $65 = 25 + 4 \cdot 10$.

An explicit formula for $\begin{bmatrix} p \\ r \end{bmatrix}$ in terms of binomial coefficients is

$$\begin{bmatrix} p \\ r \end{bmatrix} = \begin{cases} \binom{r}{0} r^p - \binom{r}{1}(r - 1)^p + \cdots + (-1)^{r-1} \binom{r}{r - 1} 1^p \\ \qquad\qquad\qquad\qquad \text{for } 1 \leqq r \leqq p \\ 0 \qquad\qquad\qquad\qquad \text{for } r > p. \end{cases}$$

This can be proved by mathematical induction, using equation (2). We shall omit the details.

To complete the solution to the problem we have merely to observe that $f(p,r) = \begin{bmatrix} p \\ r \end{bmatrix} m(m-1) \cdots (m-r+1)$, so that the required probability is

$$\frac{\begin{bmatrix} p \\ r \end{bmatrix} m(m-1) \cdots (m-r+1)}{m^p}.$$

Remark. From the second solution to b one can obtain a new solution to part a, since a is the special case of b where $r = m$.

81c. If there are fewer passengers than carriages, then the probability that every carriage will be occupied is 0. Therefore the solution to part a must vanish when $p < m$, that is

$$m^p - \binom{m}{1}(m-1)^p + \binom{m}{2}(m-2)^p - \cdots + (-1)^{m-1}\binom{m}{m-1}1^p = 0$$

$$\text{for} \quad p < m.$$

Since $\binom{m}{s} = \binom{m}{m-s}$, this can be written in the form

$$\binom{m}{1}1^p - \binom{m}{2}2^p + \binom{m}{3}3^p - \cdots + (-1)^{m-1}\binom{m}{m}m^p = 0 \quad \text{for} \quad p < m.$$

Next suppose $p = m$, i.e., the number of passengers is equal to the number of carriages. Then the only arrangements in which all carriages are occupied are those where there is exactly one person in each carriage. There are $m!$ such arrangements, and so the probability of such an arrangement is $m!/m^m$. Comparing this with the solution found in part a, we see that

$$m^m - \binom{m}{1}(m-1)^m + \binom{m}{2}(m-2)^m - \cdots + (-1)^{m-1}\binom{m}{m-1}1^m = m!$$

Multiplying by $(-1)^{m-1}$ and using the fact that $\binom{m}{s} = \binom{m}{m-s}$, this becomes

$$\binom{m}{1}1^m - \binom{m}{2}2^m + \binom{m}{3}3^m - \cdots + (-1)^{m-1}\binom{m}{m}m^m = (-1)^{m-1}m!$$

82. Let us first compute the total number of possible outcomes of the experiment, i.e. the number of ways in which one can arrange the 20 slips in a circle so that capital letters alternate with small letters.

In doing this we will regard two arrangements as the same if they can be obtained from each other by rotating the circle. With this convention we may suppose that the letter A is always at the top of the circle, since each arrangement can be rotated until this is the case. There are then 9 positions for the remaining capital letters, and so these can be arranged in 9! ways. The small letters can be arranged in 10! ways, and hence the total number of possible outcomes is 9! 10!

Let us now compute the number of *unfavorable* outcomes, i.e. those in which at least one capital letter is next to the corresponding small letter. Let A_1 be the set of outcomes in which A is next to a, A_2 those in which B is next to b, . . . , and A_{10} those in which J is next to j.

Then $A_1 \cup A_2 \cup \cdots \cup A_{10}$ is the set of all unfavorable outcomes. We can calculate $\#(A_1 \cup A_2 \cup \cdots \cup A_{10})$ by the principle of inclusion and exclusion, provided we know the quantities $\#(A_i)$, $\#(A_i \cap A_j)$, $\#(A_i \cap A_j \cap A_k)$, etc. By symmetry all the terms $\#(A_i)$ have the same value, which we will call a_1. Similarly all the terms $\#(A_i \cap A_j)$ have the same value, which we will call a_2. We similarly define $a_3, a_4, \ldots, a_{10}$. Then the principle of inclusion and exclusion gives

$$\#(A_1 \cup \cdots \cup A_{10}) = 10a_1 - \binom{10}{2}a_2 + \binom{10}{3}a_3 - \cdots - \binom{10}{10}a_{10}.$$

It remains only to determine the numbers $a_1, a_2, \ldots, a_{10}$.

The number a_1 of arrangements in which a given capital letter and the corresponding small letter (say, A and a) are next to each other can be determined in the following way. Arrange the 18 slips bearing the letters other than A and a in a circle in any way such that capital letters alternate with small letters; this can be done in $9! \cdot 8!$ different ways (compare with the proof that the total number of all possible arrangements is $10! \cdot 9!$). Then insert the slips bearing the letters A and a between any two of the other letters (A is, of course, to be put next to a small letter and a next to a capital letter); this can be done in 18 different ways, since there are 18 different places where the two extra letters can be inserted. It follows that

$$a_1 = 9! \cdot 8! \cdot 18; \quad \binom{10}{1}a_1 = 10! \cdot 9! \cdot 2.$$

Similarly, to determine a_2 we arrange the 16 letters exlcusive of A, a, B, and b in a circle; this can be done in $8! \cdot 7!$ ways. Then insert A and a between two consecutive letters (with A next to the small letter and a next to the capital letter); this can be done in 16 ways. Finally insert B and b between any two consecutive letters other than A and a (which must not be separated); this can be done in 17 different ways (there are 18 pairs of adjacent letters, between all but one of which B and b can be inserted). It follows from these considerations that

$$a_2 = 8! \cdot 7! \cdot 16 \cdot 17; \quad \binom{10}{2}a_2 = \frac{10 \cdot 9}{1 \cdot 2} 8! \cdot 7! \cdot 16 \cdot 17 = 10! \cdot 8! \cdot 17.$$

All the numbers a_k, where $k = 1, 2, \ldots, 9$, can be determined in this way. To find the number a_k, we arrange in a circle the $20 - 2k$ letters left after deleting k capital letters and the corresponding k small letters; this can be done in $(10 - k)! \, (10 - k - 1)!$ ways. We then insert the k pairs of capital and small letters between the other letters, which can be done in $(20 - 2k)(20 - 2k + 1) \cdots (20 - 2k + k - 1)$ ways. (Inserting a capital and its corresponding small letter increases by 2 the total number of letters but only increases by 1 the number of places where the next pair can be inserted because the preceding pair cannot be separated.) We thus obtain:

$$a_k = (10 - k)! \, (10 - k - 1)! \, (20 - 2k)(20 - 2k + 1) \cdots (20 - k - 1).$$

Hence

$$a_3 = 7! \cdot 6! \cdot 14 \cdot 15 \cdot 16;$$

$$\binom{10}{3} a_3 = \frac{10 \cdot 9 \cdot 8}{1 \cdot 2 \cdot 3} \cdot 7! \cdot 6! \cdot 14 \cdot 15 \cdot 16 = 10! \cdot 8! \cdot 10;$$

$$a_4 = 6! \cdot 5! \cdot 12 \cdot 13 \cdot 14 \cdot 15;$$

$$\binom{10}{4} a_4 = \frac{10 \cdot 9 \cdot 8 \cdot 7}{1 \cdot 2 \cdot 3 \cdot 4} 6! \cdot 5! \cdot 12 \cdot 13 \cdot 14 \cdot 15 = 10! \cdot 7! \cdot \frac{65}{2};$$

$$a_5 = 5! \cdot 4! \cdot 10 \cdot 11 \cdot 12 \cdot 13 \cdot 14;$$

$$\binom{10}{5} a_5 = \frac{10 \cdot 9 \cdot 8 \cdot 7 \cdot 6}{1 \cdot 2 \cdot 3 \cdot 4 \cdot 5} 5! \cdot 4! \cdot 10 \cdot 11 \cdot 12 \cdot 13 \cdot 14 = 10! \cdot 7! \frac{143}{15};$$

$$a_6 = 4! \cdot 3! \cdot 8 \cdot 9 \cdot 10 \cdot 11 \cdot 12 \cdot 13;$$

$$\binom{10}{6} a_6 = \frac{10 \cdot 9 \cdot 8 \cdot 7 \cdot 6 \cdot 5}{1 \cdot 2 \cdot 3 \cdot 4 \cdot 5 \cdot 6} 4! \cdot 3! \cdot 8 \cdot 9 \cdot 10 \cdot 11 \cdot 12 \cdot 13 = 10! \cdot 4! \cdot 429;$$

$$a_7 = 3! \cdot 2! \cdot 6 \cdot 7 \cdot 8 \cdot 9 \cdot 10 \cdot 11 \cdot 12;$$

$$\binom{10}{7} a_7 = \frac{10 \cdot 9 \cdot 8 \cdot 7 \cdot 6 \cdot 5 \cdot 4}{1 \cdot 2 \cdot 3 \cdot 4 \cdot 5 \cdot 6 \cdot 7} 3! \cdot 2! \cdot 6 \cdot 7 \cdot 8 \cdot 9 \cdot 10 \cdot 11 \cdot 12 = 10! \cdot 4! \cdot 66;$$

$$a_8 = 2! \cdot 1! \cdot 4 \cdot 5 \cdot 6 \cdot 7 \cdot 8 \cdot 9 \cdot 10 \cdot 11;$$

$$\binom{10}{8} a_8 = \frac{10 \cdot 9 \cdot 8 \cdot 7 \cdot 6 \cdot 5 \cdot 4 \cdot 3}{1 \cdot 2 \cdot 3 \cdot 4 \cdot 5 \cdot 6 \cdot 7 \cdot 8} 2! \cdot 1! \cdot 4 \cdot 5 \cdot 6 \cdot 7 \cdot 8 \cdot 9 \cdot 10 \cdot 11$$
$$= 10! \cdot 165;$$

$$a_9 = 1! \cdot 1 \cdot 2 \cdot 3 \cdot 4 \cdot 5 \cdot 6 \cdot 7 \cdot 8 \cdot 9 \cdot 10;$$

$$\binom{10}{9} a_9 = \frac{10 \cdot 9 \cdot 8 \cdot 7 \cdot 6 \cdot 5 \cdot 4 \cdot 3 \cdot 2}{1 \cdot 2 \cdot 3 \cdot 4 \cdot 5 \cdot 6 \cdot 7 \cdot 8 \cdot 9} 1! \cdot 1 \cdot 2 \cdot 3 \cdot 4 \cdot 5 \cdot 6 \cdot 7 \cdot 8 \cdot 9 \cdot 10$$
$$= 10! \cdot 10.$$

To determine a_{10}, the number of arrangements in which each capital letter is placed next to the corresponding small letter, we proceed as follows. Arrange the 10 capital letters in a circle; as was shown above, this can be done in 9! essentially different ways. The 10 small letters can then be placed next to the corresponding capital letters in two ways (each small letter can be to the left of the corresponding capital letter or to the right). Thus,

$$a_{10} = 2 \cdot 9!; \quad \binom{10}{10} a_{10} = 2 \cdot 9!.$$

Finally, for the number of favorable outcomes we obtain:

$$10! \cdot 9! - \#(A_1 \cup \cdots \cup A_{10})$$

$$= 10! \cdot 9! - 10! \cdot 9! \cdot 2 + 10! \cdot 8! \cdot 17$$

$$- 10! \cdot 8! \cdot 10 + 10! \cdot 7! \frac{65}{2} - 10! \cdot 7! \cdot \frac{143}{15}$$

$$+ 10! \cdot 4! \cdot 429 - 10! \cdot 4! \cdot 66 + 10! \cdot 165$$

$$- 10! \cdot 10 + 2 \cdot 9!$$

$$= 10!\left(-9! + 8! \cdot 7 + 7! \cdot \frac{689}{30} + {}_4! \cdot 363 + 155\right) + 2 \cdot 9!$$

$$= 9! \cdot 439{,}792,$$

which means that the required probability is

$$\frac{9! \, 439{,}792}{10! \, 9!} = \frac{439{,}792}{10!} = \frac{439{,}792}{3{,}528{,}800} \approx 0.12.$$

83a. *First solution.* The experiment under consideration here consists of $n + m$ customers, n of whom have five-dollar bills and the other m of whom have ten-dollar bills, getting in line to buy tickets. The total number of possible outcomes to this experiment is equal to the number of ways the m customers with only ten-dollar bills can be arranged in the line of $n + m$ people, that is, $\binom{n+m}{m}$. We will represent these $\binom{n+m}{m}$ possibilities with the aid of the $\binom{n+m}{m}$ shortest paths joining the intersections $(0,0)$ and (n,m) of a network of roads such as we considered above (see the remarks preceding problem 61). Specifically, starting from the point $A_0 = (0,0)$ lay off a segment A_0A_1 of length 1 either to the right (if the first customer has a five-dollar bill) or upwards (if he has only a ten-dollar bill). From the point A_1, lay off a segment A_1A_2 of length 1 either to the right or upwards, according to whether the second customer has a five-dollar bill or only has a ten-dollar bill. From the point A_2 lay off a segment A_2A_3 or length 1, horizontally or vertically according to whether the third customer has a five-dollar bill or has only a ten-dollar bill, etc. (fig. 61a.). Thus each of the $\binom{m+n}{m}$ possible arrangements of the $n+m$ customers into a line corresponds to a path $A_0A_1A_2\cdots A_{n+m}$ consisting of n horizontal segments and m vertical segments. All these paths end at the point $A_{n+m}=(n,m)$ which lies n units to the right of and m units above the point A_0; they constitute all possible shortest paths joining the intersections $A_0=(0,0)$ and $A_{n+m}=(n,m)$.

Let us find the number of cases in which none of the customers has to wait for change. In order for this to happen, it is necessary and sufficient that in front of each customer there be at least as many customers with five-dollar bills as customers who have only tens. This means geometrically

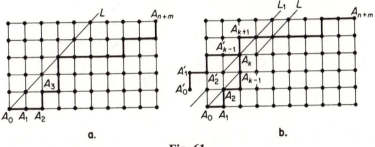

Fig. 61

that every favorable outcome corresponds to a path $A_0A_1A_2\cdots A_{n+m}$ which lies entirely below the straight line L which passes through the point A_0 at an angle of $45°$ above the horizontal (fig. 61a); in particular, the first segment of such a path must be horizontal.

It follows from this that every path corresponding to an unfavorable outcome must cross the line L, or what is the same thing, must have a vertex lying on the line L_1 which is parallel to L and obtained by moving L one unit upwards (fig. 61b). For $m>n$ our paths will necessarily have a vertex on the line L_1; for in this case the point A_{n+m} is located above the line L. Suppose now that $m\leq n$; let us find how many of the paths have a vertex on the line L_1. Let $A_0A_1A_2\cdots A_{n+m}$ be such a path and A_k the first of its vertices which lies on the line L_1. Reflect the portion $A_0A_1\cdots A_k$ of this path over the line L_1. We obtain a path $A_0'A_1'\cdots A_{k-1}'A_kA_{k+1}\cdots A_{n+m}$ which joins A_{n+m} to the point A_0' which is symmetric about L_1 to the point A_0 (that is, it is located one unit above and one unit to the left of A_0; see fig. 61b). Further, for $m\leq n$, every shortest path joining the points A_0' and A_{n+m} necessarily intersects the line L_1. If A_k is the first point of intersection of any path $A_0'A_1' \cdots A_{k-1}A_kA_{k+1} \cdots A_{n+m}$ with the line L_1, then by reflecting the portion $A_0'A_1' \cdots A_{k-1}'A_k$ over L_1 we obtain a path $A_0A_1 \cdots A_kA_{k+1} \cdots A_{n+m}$ joining A_0 to A_{n+m} and having a vertex on the line L_1. Thus for $m \leq n$ the number of paths joining A_0 to A_{n+m} and having a vertex on the line L_1 coincides with the number of paths joining A_0' to A_{n+m}. But these latter paths each consist of $n + 1$ horizontal segments and $m - 1$ vertical segments; there are hence $\binom{n+m}{m-1}$ of them.

Thus in the problem under consideration there are a total of $\binom{n+m}{m}$ (equally likely) possible outcomes; for $m > n$ the number of unfavorable outcomes equals the total number of outcomes, and for $m \leq n$ it equals $\binom{n+m}{m-1}$. It follows from this that the number of favorable outcomes is zero when $m > n$, and is

$$\binom{n+m}{m} - \binom{n+m}{m-1} = \frac{(n+m)!}{n!\,m!} - \frac{(n+m)!}{(n+1)!\,(m-1)!}$$

$$= \frac{(n+m)!}{n!\,(m-1)!}\left\{\frac{1}{m} - \frac{1}{n+1}\right\}$$

$$= \frac{(n+m)!\,(n-m+1)}{(n+1)!\,m!} \quad \text{when } m \leq n.$$

This means that the probability that none of the customers has to wait for change is 0 for $m > n$, and for $m \leq n$ is equal to

$$\frac{(n+m)!\,(n-m+1)}{(n+1)!\,m!} \bigg/ \binom{n+m}{m} = \frac{(n+m)!\,(n-m+1)}{(n+1)!\,m!} \bigg/ \frac{(n+m)!}{n!\,m!}$$

$$= \frac{n-m+1}{n+1}.$$

Second solution. In this section we will discuss a method which does not *derive* the formula for the number of favorable outcomes, but merely proves its correctness once it has been guessed. Suppose $n > 0$, and let $S(n,m)$ be the number of favorable outcomes. Then, as in part a, $S(n,m)$ is the number of paths with n horizontal and m vertical segments joining the point $(0,0)$ to the point (n,m), and having no vertices on the line L_1 of fig. 61b. We have $S(n,m) = 0$ when $n < m$, since then the point (n,m) lies above L.

Also, we have $S(n,0) = 1$, since the only path with n horizontal segments and no vertical segments is a horizontal line. The next step is to derive a recursion formula for $S(n,m)$ when $n,m > 0$. To do this, note that the paths ending in (n,m) are of two types: those which pass through

$(n - 1, m)$ and those which pass through $(n, m - 1)$. There are $S(n - 1, m)$ paths of the first type and $S(n, m - 1)$ of the second type. Therefore

$$S(n,m) = S(n - 1, m) + S(n, m - 1). \tag{1}$$

Now suppose we have guessed that when $0 \leqq m \leqq n + 1$,

$$S(n,m) = \frac{n - m + 1}{n + 1}\binom{n + m}{m} \tag{2}$$

(say on the basis of a direct computation of $S(n,m)$ for small values of n and m). We can then prove our conjecture by mathematical induction on the sum $n + m$. When $n + m = 1$ one can verify (2) directly. Suppose we have already shown that

$$S(a,b) = \frac{a - b + 1}{a + b}\binom{a + b}{b}$$

for all pairs a,b such that $0 \leqq b \leqq a + 1$ and $a + b < n + m$. We must then derive (2) from this assumption. If $m = 0$, then (2) reduces to $S(n,0) = 1$, which is true by a remark made above. If $m = n + 1$, it reduces to $S(n, n + 1) = 0$ which was also noted earlier. When $0 < m < n + 1$, we use equation (1), and apply the induction hypothesis to the two terms on the right-hand side. This gives

$$\begin{aligned}
S(n,m) &= \frac{(n - 1) - m + 1}{(n - 1) + 1}\binom{n - 1 + m}{m} \\
&\quad + \frac{n - (m - 1) + 1}{n + 1}\binom{n + m - 1}{m - 1} \\
&= \frac{n - m}{n}\frac{(n - m - 1)!}{(n - 1)!\,m!} + \frac{n - m + 2}{n + 1}\frac{(n - m - 1)!}{n!\,(m - 1)!} \\
&= \frac{(n - m - 1)!}{n!\,(m - 1)!}\left(\frac{n - m}{m} + \frac{n - m + 2}{n + 1}\right) \\
&= \frac{(n - m - 1)!}{n!\,(m - 1)!}\frac{(n + m)(n - m + 1)}{(n + 1)m} \\
&= \frac{n - m + 1}{n + 1}\binom{n + m}{m}.
\end{aligned}$$

This completes the induction, and shows that (2) is valid for all m, n with $0 \leqq m \leqq n + 1$. The required probability is then

$$\frac{S(n,m)}{\binom{n + m}{m}} = \begin{cases} \dfrac{n - m + 1}{n + 1} & \text{for } 0 \leqq m \leqq n + 1 \\[2mm] 0 & \text{otherwise.} \end{cases}$$

Third solution.[6]   We have to find the probability $p(n,m)$ of the event A that at each point in the ticket line there are at least as many customers with five-dollar bills as there are with ten-dollar bills ahead of that point. Let $q(n,m)$ be the probability of the event B that at each point of the line (including the end) there are *more* customers with five-dollar bills than there are with ten-dollar bills ahead of that point. In order for event B to occur, the first customer must have a five-dollar bill, and the remaining $n + m - 1$ customers must be arranged so that among themselves (i.e., without the first customer), they satisfy the event A. Since the probability that the first customer has a five-dollar bill is $n/(n + m)$, we therefore have

$$q(n,m) = \frac{n}{n + m}\, p(n - 1, m).$$

If we can determine $q(n,m)$, we will then be able to find $p(n,m)$ by using this formula. So we will now show how to determine $q(n,m)$.

It is clear that for $n \leq m$, $q(n,m) = 0$. Suppose now that $n > m$ and consider any arrangement of the $n + m$ customers (n of whom have five-dollar bills and the other m of whom have only tens) into a line. From this arrangement we can obtain $n + m - 1$ new arrangements as follows: move the first customer from the front of the line to the last place (thus creating a new arrangement in which the 2nd, 3rd, . . . , and $(n + m)$th customers of the original one are one place further forward); then repeat this process (thus giving a line in which the first two people of the original line occupy the two last places and the other people are two places further forward than originally). By repeating this process until the original arrangement is finally recovered, we obtain a total of $n + m - 1$ new arrangements. Adding to these the original arrangement, we obtain a total of $n + m$ arrangements; we will now show that exactly $n - m$ of them are favorable, i.e. are such that event B takes place.

Let us make use of the same geometric representation of the different arrangements of the $n + m$ people in line as in the first solution to this problem (compare with pp. 168–169 and fig. 61a). To every arrangement there will correspond a path $A_0 A_1 A_2 \cdots A_{n+m-1} A_{n+m}$ consisting of n horizontal segments and m vertical segments (fig. 62a). To the favorable arrangements there will correspond paths in which every vertex other than A_0 is preceded by more horizontal segments than vertical ones, that is, paths which (1) lie below the line L which passes through the point A_0 at an angle of 45° above the horizontal and (2) have no vertices other than A_0 on the line L. It is convenient to imagine the path $A_0 A_1 A_2 \cdots A_{n+m}$ as a staircase leading from the point A_0 to the point A_{n+m} and consisting of a certain number of steps of various heights and widths (where the sum

---

[6] This solution is somewhat longer than the preceding ones, but it has the advantage that it can be easily generalized. (See for example the first solution to part c.)

of the widths of all the steps is n and the sum of the heights is m). If we illuminate this staircase from above with a beam of parallel rays which make an angle of $45°$ with the horizontal, then the favorable arrangements of the customers will be those corresponding to staircases in which light falls on the base A_0 (that is, A_0 does not lie in the shadow thrown by the steps of the staircase).

To represent all $n + m$ arrangements obtained from a given one by successively moving the first customer to the end of the line, adjoin to the end A_{n+m} of the path $A_0A_1A_2 \cdots A_{n+m}$ a path $A_{n+m}A_{n+m+1}A_{n+m+2} \cdots A_{2n+2m}$ which is an exact replica of $A_0A_1A_2 \cdots A_{n+m}$ (fig. 62b). Then our $n + m$ arrangements will correspond to the paths which consist of $n + m$ consecutive unit segments and start respectively at the points A_0, A_1,

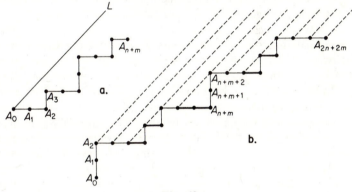

Fig. 62

$A_2, \ldots, A_{n+m-1}$ (and consequently end respectively at the points A_{n+m}, $A_{n+m+1}, A_{n+m+2}, \ldots, A_{2n+2m-1}$). If we now illuminate the path $A_0A_1A_2 \cdots A_{2n+2m}$ from above with a beam of parallel rays falling at an angle of $45°$ to the horizontal, then the favorable arrangements of customers will correspond to paths $A_kA_{k+1}A_{k+2} \cdots A_{n+m+k}$ which start at a point A_k on which light falls.[7] Thus we have only to compute how many of the points $A_0, A_1, A_2, \ldots, A_{n+m-1}$ are illuminated by the parallel beam.

Of the points $A_0, A_1, \ldots, A_{n+m-1}$, the only ones which have a chance of being illuminated are the n points A_i for which the segment A_iA_{i+1} is horizontal. But even such a point A_i need not be illuminated, for the segment A_iA_{i+1} may lie in the shade cast by a later vertical segment. Denote by v the number of vertical unit segments of the path $A_0, A_1, \ldots,$

[7] If the point A_k does not lie in the shade of the path $A_kA_{k+1}A_{k+2} \cdots A_{n+m+k}$, then A_k cannot lie in the shade thrown by the steps from A_{n+m+k} to A_{2n+2m} either. This follows from the fact that the path $A_{n+m+k}A_{n+m+k+1} \cdots A_{2n+2m}$ is a duplicate of $A_kA_{k+1} \cdots A_{n+m}$.

A_{n+m} which cast their shadows to the left of A_0 (i.e. segments like A_0A_1 and A_1A_2 in fig. 62b). Then the other $m - v$ vertical segments of $A_0A_1 \cdots A_{n+m}$ cast their shadows on horizontal segments. Since $A_{n+m}A_{n+m+1} \cdots A_{2n+2m}$ is a replica of $A_0A_1 \cdots A_{n+m}$, exactly v vertical segments of $A_{n+m}A_{n+m+1} \cdots A_{2n+2m}$ cast their shadows to the left of A_{n+m}. These shadows fall on horizontal segments of $A_0A_1 \cdots A_{n+m}$; hence there are altogether $(m - v) + v = m$ horizontal segments of $A_0A_1 \cdots A_{n+m}$ in the shade. Thus, of the n points A_i which had a chance of being illuminated, exactly m are eliminated by virtue of the shadow cast on the segment A_iA_{i+1}. Hence there remain $n - m$ illuminated points among $A_0, A_1, \ldots, A_{n+m-1}$. For example, in fig. 62b, $n = 7, m = 4$, and exactly $7 - 4 = 3$ of the points are illuminated. We have thus proved that for $n > m$ exactly $n - m$ of the $n + m$ arrangements of customers obtained from any one arrangement by successively putting the first customer at the end of the line have more people with five-dollar bills than with tens in front of each customer.

Note that the $n + m$ arrangements obtained from a single one by successively moving the first customer to the end of the line are not necessarily all distinct. If n and m are not relatively prime, then it may happen that a line of $n + m$ people consists of several parts which are exact repetitions of each other (for example, six customers with fives and three customers with only tens can be arranged in the following order: 5 10 5 5 10 5 5 10 5; the numbers 5 and 10 here denote respectively customers having fives and customers having only tens). In this case we can move the first customer to the end of the line less than $n + m$ times (three times in our example) and arrive at an arrangement identical to the initial one. However, it is easy to see that in this case each different arrangement will be repeated the same number of times (in our example $n + m = 9$ and among the nine arrangements, there are three different ones, each repeated three times).

Therefore, the ratio of the number of *different* favorable arrangements to the total number of *different* arrangements is the same as the ratio which we obtain by failing to identify identical arrangements, that is, it equals $(n - m)/(n + m)$. Thus, for $n > m$ all possible arrangements of the $n + m$ customers into a line can be divided into groups, in each of which the ratio of the number of favorable arrangements to the total number of arrangements equals $(n - m)/(n + m)$.

We have therefore shown that

$$q(n,m) = \begin{cases} 0 & \text{if } n \leqq m \\ \dfrac{n - m}{n + m} & \text{if } n > m. \end{cases}$$

We saw earlier that

$$p(n-1, m) = \frac{n+m}{n} q(n,m)$$

$$= \begin{cases} 0 & \text{if} \quad n \leq m \\ \dfrac{n-m}{n} & \text{if} \quad n > m. \end{cases}$$

Replacing n by $n+1$, we get

$$p(n,m) = \begin{cases} 0 & \text{if} \quad n+1 \leq m \\ \dfrac{n+1-m}{n+1} & \text{if} \quad n+1 > m. \end{cases}$$

Remark. For the special case $n = m$, a fourth solution is indicated in the remark at the end of the solution to problem 84a.

83b. *First solution.* Part b can be solved similarly to the first solution to part a. The initial presence of p five-dollar bills in the till means that it is possible to give immediate change to every customer if and only if for each customer in the line the number of people in front of him who have only ten-dollar bills does not exceed by more than p the number of people who have fives. Geometrically, this means that the paths corresponding to favorable outcomes lie entirely below the line L_p, which is obtained by moving the line L p units upwards (fig. 63). In other words, the paths corresponding to unfavorable outcomes are those which have points lying on the line L_{p+1} obtained by moving line L_p one unit upwards, (see fig. 63,

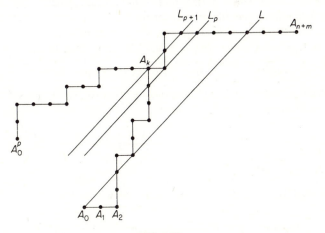

Fig. 63

where $p = 3$). It is clear that for $m > n + p$, all our paths will have vertices on the line L_{p+1}; therefore in this case the required probability is zero. On the other hand, for $m \leq p$ all outcomes are favorable and the required probability is 1.

We will now assume that $m \leq n + p$ but $m \geq p + 1$. It can be proved by the same method as in the first solution to problem 83a that the number of paths corresponding to unfavorable outcomes is in this case equal to the number of paths joining the point A_{n+m} to the point $A_0{}^p$ obtained by reflecting A_0 in the line L_{p+1} (that is, the point $p + 1$ units above and $p + 1$ units to the left of A_0). It follows from this that the total number of unfavorable outcomes is $\binom{n+m}{m-p-1}$, whence the total number of favorable outcomes is $\binom{n+m}{m} - \binom{n+m}{m-p-1}$. Thus, for $n + p \geq m \geq p + 1$ the probability that none of the customers will have to wait for change is

$$\frac{\binom{n+m}{m} - \binom{n+m}{m-p-1}}{\binom{n+m}{m}} = 1 - \frac{(m+n)!}{(m-p-1)!\,(n+p+1)!} \Big/ \frac{(m+n)!}{m!\,n!}$$

$$= 1 - \frac{m(m-1)\cdots(m-p)}{(n+1)(n+2)\cdots(n+p+1)}.$$

Second solution.   If we have been able to arrive at the correct formula by some process of educated guesswork, we can prove it by mathematical induction as in the second solution of part a. For let $S_p(n,m)$ be the number of paths with n horizontal and m vertical segments, leading from $(0,0)$ to (n,m), and having no vertices on L_{p+1}. We see as in part a that if $m,n > 0$, then $S_p(n,m) = S_p(n-1,m) + S_p(n,m-1)$. If $m > n + p$, then $S_p(n,m) = 0$, since the point (n,m) lies above the line L_p. If $m \leq p$, then $S_p(n,m) = \binom{n+m}{m}$, since in this case no path from $(0,0)$ to (n,m) can have a vertex on L_{p+1}. Suppose we have somehow guessed that for $p \leq m \leq n + p + 1$,

$$S_p(n,m) = \left[1 - \frac{m(m-1)\cdots(m-p)}{(n+1)(n+2)\cdots(n+p+1)}\right]\binom{n+m}{m}. \quad (1)$$

We can then prove (1) by induction on $n + m$. When $n + m = 1$, (1) can be verified directly. Assume we already know that

$$S_p(a,b) = \left[1 - \frac{b(b-1)\cdots(b-p)}{(a+1)(a+2)\cdots(a+p+1)}\right]\binom{a+b}{b}$$

for all pairs a, b with $p \leq b \leq a + p + 1$ and $a + b < n + m$. If $m = p$, then (1) reduces to $S_p(n,p) = \binom{n+p}{p}$, which is true as noted above. When $m = n + p + 1$, (1) reduces to $S_p(n, n + p + 1) = 0$, which we also know to be correct from the above remarks. If $p < m < n + p + 1$, we use the equation $S_p(n,m) = S_p(n, m - 1) + S_p(n - 1, m)$, applying the induction hypothesis to both terms on the right. This gives

$$
\begin{aligned}
S_p(n,m) &= \left[1 - \frac{(m-1)(m-2)\cdots(m-p-1)}{(n+1)(n+2)\cdots(n+p+1)} \right]\binom{n+m-1}{m-1} \\
&\quad + \left[1 - \frac{m(m-1)\cdots(m-p)}{n(n+1)\cdots(n+p)} \right]\binom{n+m-1}{m} \\
&= \frac{(n+m-1)!}{n!\,(m-1)!} + \frac{(n+m-1)!}{(n-1)!\,m!} \\
&\quad - \frac{(m-1)(m-2)\cdots(m-p-1)}{(n+1)(n+2)\cdots(n+p+1)}\frac{(n-m-1)!}{n!\,(m-1)!} \\
&\quad - \frac{m(m-1)\cdots(m-p)}{n(n+1)\cdots(n+p)}\frac{(n-m-1)!}{(n-1)!\,m!} \\
&= \frac{(n+m-1)!}{(n-1)!\,(m-1)!}\left(\frac{1}{n} + \frac{1}{m}\right) \\
&\quad - \frac{(m-1)(m-2)\cdots(m-p)}{(n+1)(n+2)\cdots(n+p)}\frac{(n+m-1)!}{n!\,(m-1)!} \\
&\quad \times \left(\frac{m-p-1}{n+p+1} - 1\right) \\
&= \frac{(n+m)!}{n!\,m!} - \frac{(m-1)(m-2)\cdots(m-p)}{(n+1)(n+2)\cdots(n+p)} \\
&\quad \times \frac{(n+m-1)!}{n!\,(m-1)!}\frac{n+m}{n+p+1} \\
&= \left[1 - \frac{m(m-1)\cdots(m-p)}{(n+1)(n+2)\cdots(n+p+1)} \right]\binom{n+m}{m}.
\end{aligned}
$$

This completes the induction. The required probability is

$$
S_p(n,m) \Big/ \binom{n+m}{m}.
$$

Remark. In the special case $p = 1$ the problem can be solved by a method similar to that used in the third solution of part a. Suppose that originally there is one five-dollar bill in the cash register, and let $r(n,m)$ be the probability of the event C that the cashier can always give change when required. In order for event A to occur, the first customer must have a five-dollar bill, and the remaining $n + m - 1$ customers must be arranged so that among themselves (i.e., excluding the first customer) they satisfy the conditions for event C. Hence if $n > 0$, we have

$$p(n,m) = \frac{n}{n + m} r(n - 1, m).$$

Since

$$p(n,m) = \begin{cases} 0 & \text{if } m > n \\ \dfrac{n - m + 1}{n + 1} & \text{if } m \leqq n \end{cases}$$

we obtain

$$r(n - 1, m) = \begin{cases} 0 & \text{if } m > n \\ \dfrac{(n + m)(n - m + 1)}{n(n + 1)} & \text{if } m \leqq n. \end{cases}$$

Replacing n by $n + 1$, this becomes

$$r(n, m) = \begin{cases} 0 & \text{if } m > n + 1 \\ \dfrac{(n + m + 1)(n - m + 2)}{(n + 1)(n + 2)} & \text{if } m \leqq n + 1. \end{cases}$$

83c. *First solution.* In order that none of the customers has to wait for change, it is necessary and sufficient that the number of people with one-dollar bills in front of each customer be *at least twice* the number of people in front of him who have only three-dollar bills. The simplest way of computing the probability of this event is with the aid of the method outlined in the third solution to part a; this solution carries over to part c almost word for word. Let us compute first of all the probability that at each point in the line (including the end) there are *more* than twice as many people with one-dollar bills as with three-dollar bills ahead of that point. To each such arrangement of customers there corresponds a "staircase" $A_0 A_1 A_2 \cdots A_{n+m}$ whose base A_0 will not lie in the shade when the staircase is illuminated from above by a beam of parallel rays which fall at an angle such that the shade thrown by any vertical segment is twice the length of that segment (in other words, the tangent of the angle between the rays and the horizontal is $1/2$; see fig. 64). As in the third solution to part a, it can be shown that the probability that in front of each customer there are more than twice as many customers with dollars as with three-dollar bills is $(n - 2m)/(n + 2m)$ (we assume here that $n > 2m$; for $n \leqq 2m$ the probability in question is zero).

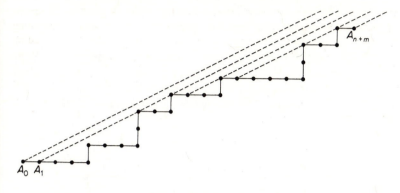

Fig. 64

Further, reasoning as in the third solution to problem 83a, we arrive at the following conclusion:

Let there be two lines, the first consisting of n customers who have one-dollar bills and m who have only three-dollar bills (a total of $n + m$ people), and the second consisting of $n + 1$ customers with one-dollar bills and m with only three-dollar bills (a total of $n + m + 1$ people). Then the probability that in the first line there are *at least* as many people with dollars as people with only threes in front of a given customer is $(n + m + 1)/(n + 1)$ times the probability that in the second line there are *more* than twice as many people with dollars as people with only threes in front of the customer.

It follows from this that the probability to be computed in this problem is

$$\frac{n - 2m + 1}{n + m + 1} \cdot \frac{n + m + 1}{n + 1} = \frac{n - 2m + 1}{n + 1}$$

for $n \geqq 2m$, and zero for $n < 2m$.

Second solution. The problem also has a solution whose general idea is close to that of the first solution of part a, but is appreciably more complicated.

As in the first solution to part a, we represent the $\binom{n + m}{m}$ equally likely arrangements of the $n + m$ customers into line with the aid of the $\binom{n + m}{m}$ shortest paths joining the points

$$A_0 = (0,0) \quad \text{and} \quad A_{n+m} = (n,m).$$

Then the favorable arrangements are those which correspond to paths located on or below the line $\tilde{L}$ which passes through the points

$$A_0, B_1 = (2,1) \ B_2 = (4,2), \ldots, B_m = (2m,m)$$

(fig. 65a).[8] For $n < 2m$ there are no such paths (for in this case the point A_{n+m} is located above the line $\tilde{L}$); from here on we will assume that $n \geqq 2m$.

Let us now compute the number of unfavorable arrangements; these are the arrangements which correspond to paths from A_0 to A_{n+m} which cross the line $\tilde{L}$. All such paths will have vertices on the line $\tilde{L}_1$ which is parallel to $\tilde{L}$ and obtained by translating $\tilde{L}$ a distance of 1 unit upwards

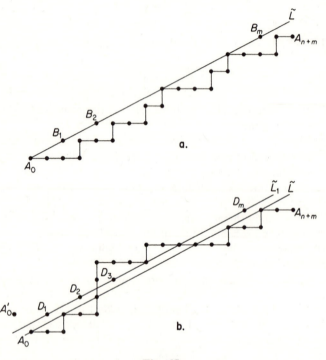

Fig. 65

(fig. 65b). We thus have only to compute how many paths from A_0 to A_{n+m} have vertices on $\tilde{L}_1$.

As in the first solution to part a, denote by A_0' the point located one unit above and one unit to the left of A_0 (fig. 65b). We will show that the number of shortest paths from A_0 to A_{n+m} which have vertices on the line $\tilde{L}_1$ is exactly twice the number of shortest paths from A_0' to A_{n+m}.

Denote by N_{AB} the number of shortest paths joining the points

---

[8] In the Cartesian coordinate system whose origin is at A_0 and whose x-axis and y-axis are respectively horizontal and vertical, the equation of the line $\tilde{L}$ can be written in the form $x = 2y$.

A and B; if $A = (i,j)$ and $B = (k,l)$, then $N_{AB} = \begin{pmatrix} k - i + l - j \\ l - j \end{pmatrix}$. The following points are located on the line $\tilde{L}_1$: $D_1 = (1,1)$, $D_2 = (3,2)$, $D_3 = (5,3)$, ... $D_k = (2k - 1, k)$, ..., $D_m = (2m - 1, m)$. Since

$$N_{A_0 D_k} = \begin{pmatrix} 3k - 1 \\ k \end{pmatrix} = \frac{(3k - 1)!}{k!\,(2k - 1)!} = 2\,\frac{(3k - 1)!}{(k - 1)!\,(2k)!} = 2\begin{pmatrix} 3k - 1 \\ k - 1 \end{pmatrix}$$

and

$$N_{A_0' D_k} = \begin{pmatrix} 3k - 1 \\ k - 1 \end{pmatrix},$$

for arbitrary k ($k = 1, 2, \ldots, m$) we have

$$N_{A_0 D_k} = 2N_{A_0' D_k}.$$

Let us now compare the number of shortest paths from A_0 to A_{n+m} which first meet $\tilde{L}_1$ at the point D_k with the number of such shortest paths from A_0' to A_{n+m}.

The number of shortest paths from A_0 to A_{n+m} which pass through the point D_1 is equal to

$$N_{A_0 D_1} \cdot N_{D_1 A_{n+m}} = 2N_{A_0' D_1} \cdot N_{D_1 A_{n+m}},$$

and the number of shortest paths from A_0' to A_{n+m} which pass through the point D_1 is $N_{A_0' D_1} \cdot N_{D_1 A_{n+m}}$. Consequently, there are exactly twice as many of the first kind of paths as of the second.

The number of shortest paths from A_0 to A_{n+m} which pass through the point D_2 but do not pass through the point D_1 is

$$N_{A_0 D_2} \cdot N_{D_2 A_{n+m}} - N_{A_0 D_1} \cdot N_{D_1 D_2} \cdot N_{D_2 A_{n+m}}$$

(the first term here gives the total number of paths from A_0 to A_{n+m} which pass through D_2 and the second gives the number of such paths which also pass through D_1). The number of such paths from A_0' to A_{n+m} is

$$N_{A_0' D_2} N_{D_2 A_{n+m}} - N_{A_0' D_1} N_{D_1 D_2} N_{D_2 A_{n+m}}.$$

Since $N_{A_0 D_2} = 2N_{A_0' D_2}$ and $N_{A_0 D_1} = 2N_{A_0' D_1}$, the number of paths from A_0 to A_{n+m} which pass through D_2 but not D_1 is exactly twice the number of such paths from A_0' to A_{n+m}.

Similarly, the number of shortest paths from A_0 to A_{n+m} which pass through D_3, but do not pass through either D_1 or D_2, is

$$N_{A_0 D_3} \cdot N_{D_3 A_{n+m}} - N_{A_0 D_1} \cdot N_{D_1 D_3} \cdot N_{D_3 A_{n+m}}$$
$$- N_{A_0 D_2} \cdot N_{D_2 D_3} \cdot N_{D_3 A_{n+m}} + N_{A_0 D_1} \cdot N_{D_1 D_2} N_{D_2 D_3} \cdot N_{D_3 A_{n+m}};$$

(the second and third terms give the number of paths from A_0 to A_{n+m} which pass through D_1 and D_2 respectively, and the fourth term gives the number of paths which pass through all three points D_1, D_2, and D_3). The number of such paths which join A_0' to A_{n+m} is

$$N_{A_0'D_3} \cdot N_{D_3A_{n+m}} - N_{A_0'D_1} \cdot N_{D_1D_3} \cdot N_{D_3A_{n+m}}$$
$$- N_{A_0'D_2} \cdot N_{D_2D_3} \cdot N_{D_3A_{n+m}} + N_{A_0'D_1} \cdot N_{D_1D_2} \cdot N_{D_2D_3} \cdot N_{D_3A_{n+m}};$$

and here it is again easy to see that the number of paths from A_0 to A_{n+m} is exactly twice the number of paths from A_0' to A_{n+m}. Continuing to reason in the same way, we can prove that for any k the number of shortest paths from A_0 to A_{n+m} which first meet the line $\tilde{L}_1$ at the point D_k is exactly twice the number of shortest paths from A_0' to A_{n+m} which first meet $\tilde{L}_1$ at that point.

Therefore, the total number of shortest paths from A_0 to A_{n+m} which have vertices on the line $\tilde{L}_1$ is exactly twice the number of shortest paths from A_0' to A_{n+m} which have vertices on that line. But any shortest path from A_0' to A_{n+m} has to have at least one vertex on $\tilde{L}_1$ (for $n \geq 2m$, the point A_0' is located on the left side of $\tilde{L}_1$ and the point A_{n+m} on the right side). Consequently, for $n \geq 2m$ the number of shortest paths from A_0 to A_{n+m} which have vertices on the line $\tilde{L}_1$ is $2N_{A_0'A_{n+m}}$.

Now it is easy to answer the question raised in the problem. For $n < 2m$ there are no favorable outcomes at all to the experiment; consequently in this case the required probability is zero (this, of course, was obvious to begin with). For $n \geq 2m$, the number of unfavorable outcomes is

$$2N_{A_0'A_{n+m}} = \binom{n+m}{m-1};$$

since the total number of (equally likely) possible outcomes to the experiment is $\binom{n+m}{m}$, the number of favorable outcomes for $n \geq 2m$ is

$$\binom{n+m}{m} - 2\binom{n+m}{m-1} = \frac{(n+m)!}{m!\,n!} - 2\frac{(n+m)!}{(m-1)!\,(n+1)!}$$
$$= \frac{(n+m)!}{(m-1)!\,n!}\left\{\frac{1}{m} - \frac{2}{n+1}\right\}$$
$$= \frac{(n+m)!\,(n-2m+1)}{m!\,(n+1)!},$$

and consequently, the probability sought is

$$\frac{(n+m)!\,(n-2m+1)}{m!\,(n+1)!} \Big/ \frac{(n+m)!}{m!\,n!} = \frac{n-2m+1}{n+1}.$$

Third solution. If we have been able to arrive at the answer to our problem by some process of educated guesswork, we can, as in parts a and b, verify it by mathematical induction. Let $T(n,m)$ be the number of shortest paths from (0,0) to (n,m) having no vertices on the line $\tilde{L}_1$ of fig. 65b. If $n < 2m$, then $T(n,m) = 0$, since in this case the point A_{n+m} lies above $\tilde{L}$. Also, if $m = 0$, then $T(n,0) = 1$, since the only path in this case is a horizontal line.

We next obtain a recursion formula for $T(n,m)$ when $n, m > 0$. The favorable paths from (0,0) to (n,m) fall into two categories, those passing through the point $(n, m - 1)$, and those passing through $(n - 1, m)$. There are $T(n, m - 1)$ paths of the first type, and $T(n - 1, m)$ of the second. Hence $T(n, m) = T(n, m - 1) + T(n - 1, m)$ when $n, m > 0$.

Now suppose we have somehow guessed that for $0 \leq 2m \leq n + 1$, the correct formula is

$$T(n,m) = \frac{n - 2m + 1}{n + 1}\binom{n + m}{m}. \tag{1}$$

Then we can prove (1) by mathematical induction on $n + m$. We first verify (1) directly for $n + m = 1$. Now assume we have already shown that

$$T(a,b) = \frac{a - 2b + 1}{a + 1}\binom{a + b}{b}$$

for all pairs a, b such that $0 \leq 2b \leq a + 1$, $a + b < n + m$. If $n = 2m - 1$, then (1) reduces to $T(2m - 1, m) = 0$, which is true by the remark made above. If $m = 0$, (1) reduces to $T(n, 0) = 1$, which was also noted above. If $0 < 2m < n + 1$, then we can apply the induction hypothesis to both terms on the right-hand side of the formula $T(n,m) = T(n, m - 1) + T(n - 1, m)$. This gives

$$\begin{aligned}
T(n,m) &= \frac{n - 2(m - 1) + 1}{n + 1}\binom{n + m - 1}{m - 1} \\
&\quad + \frac{(n - 1) - 2m + 1}{(n - 1) + 1}\binom{n + m - 1}{m} \\
&= \frac{n - 2m + 3}{n + 1}\frac{(n + m - 1)!}{n!\,(m - 1)!} + \frac{n - 2m}{n}\frac{(n + m - 1)!}{(n - 1)!\,m!} \\
&= \left(\frac{n - 2m + 3}{n + 1} + \frac{n - 2m}{m}\right)\frac{(n + m - 1)!}{n!\,(m - 1)!} \\
&= \frac{(n + m)(n - 2m + 1)}{(n + 1)m}\frac{(n + m - 1)!}{n!\,(m - 1)!} \\
&= \frac{n - 2m + 1}{n + 1}\frac{(n + m)!}{n!\,m!} \\
&= \frac{n - 2m + 1}{n + 1}\binom{n + m}{m},
\end{aligned}$$

completing the induction. The required probability is $T(n, m)\Big/\binom{n + m}{m}$.

Remark. Part c can also be formulated as follows: n objects possessing some property X and m objects possessing some property Y are arranged in a sequence in random order (thus, each of the $\binom{n + m}{m}$ possible arrangements of these $n + m$ objects is equally probable). What is the probability that at least twice as many objects with property X as objects with property Y stand in front of each object in the sequence?

In this form the problem admits a completely natural generalization: one may ask for the probability that in front of every object there are at least r times as many objects with property X as with property Y. It is easy to see that all three solutions for the special case of $r = 2$ carry over to the general case of any integer r (it is especially simple to transform the first solution of part c into a solution of the general problem). It then turns out that for arbitrary integral r the required probability is zero for $n < rm$ (which is obvious) and is $(n - rm + 1)/(n + 1)$ for $n \geq rm$. In the special case of $r = 1$ we arrive at a problem equivalent to part a.

84a. We have to compute F_n, the number of ways of dividing $2n$ points on the circumference of a circle into n pairs in such a way that the n chords formed by connecting the ends of each pair do not intersect each other. Denote the $2n$ points by $A_1, A_2, \ldots, A_{2n}$, proceeding counterclockwise around the circle. Each of the n chords is of the form $\overline{A_i A_j}$, where $i < j$. We will call A_i the *start* of the chord and A_j the *end* of the chord. Consider the first p points $A_1, A_2, \ldots, A_p$, where p is any integer in the range $1 \leq p \leq 2n$. Among these points there must be at least as many starts of chords as there are ends of chords (because if the end of a chord is one of the points $A_1, \ldots, A_p$, then the start of that chord is also one of the points $A_1, \ldots, A_p$).

Conversely, suppose that the points $A_1, \ldots, A_{2n}$ are divided into two sets S (the set of starts) and E (the set of ends) of n points each in such a way that for any $p \leq 2n$ there are at least as many starts as there are ends among the points $A_1, \ldots, A_p$. We will prove that there is then one and only one way of connecting the starts to the ends so that the resulting chords do not intersect.

To show this, consider the smallest j such that A_j is an end (we have $j \geq 2$ since A_1 is a start). Then A_j must be connected to A_{j-1}. For if it were connected to A_i where $i < j - 1$, then A_{j-1} would be the start of a chord intersecting $\overline{A_i A_j}$. Now delete the points A_{j-1} and A_j, and apply the same reasoning to the remaining points. We see that the first end among these points must be connected to the immediately preceding start. (There *is* a preceding start by the hypothesis that there are at least as many starts as there are ends among $A_1, \ldots, A_p$.) We then delete this pair and proceed in the same manner until all the chords are drawn.

We have thus shown that each set of n non-intersecting chords is completely determined by its n starting points, and that these can be prescribed subject only to the condition that there be at least as many starts as ends among the points $A_1, \ldots, A_p$ for any $p \leq 2n$. Therefore F_n is equal to the number of ways of specifying the starting points subject to this condition.

Now let us think of the starts as customers in a ticket line with only five-dollar bills, and of the ends as customers with only ten-dollar bills. Then we see that F_n is equal to the number of arrangements in a line of $2n$ customers, n of whom have five-dollar bills and n of whom have only tens, in such a way that in front of each customer there are at least as many people with fives as people with only tens. Since by problem 83a (with $m = n$), the number of arrangements of $2n$ customers (or points) which satisfy our hypothesis is $\binom{2n}{n} \Big/ (n + 1)$, we have

$$F_n = \frac{1}{n + 1} \binom{2n}{n}.$$

We have thus obtained a new solution to problem 54.

Another solution to problem 53b also follows immediately from this. For in the solution of problem 54 it was proved that T_n, the number of different ways of decomposing a convex n-gon into triangles with the aid of its diagonals, was connected with the number F_n by the relation

$$F_n = T_{n+2}.$$

This means that

$$T_n = F_{n-2} = \frac{1}{n - 1} \binom{2n - 4}{n - 2};$$

this result coincides with the answer to problem 53b obtained earlier by another method.

Remark. The argument presented here can also be applied in the reverse direction: assuming the answer to problem 54 (as obtained in the solution given on page 120) to be known, we obtain from it a new (fourth) solution to problem 83a for the special case of $m = n$.

84b. The solution of this problem is closely related to the solution of part a above; it differs only in that it is based on problem 83c rather than on problem 83a.

Denote the $3n$ points in counterclockwise order around the circle by $A_1, A_2, \ldots, A_{3n}$. Every triangle is of the form $A_i A_j A_k$, where $i < j < k$. We will call A_i the *start* of the triangle, A_j the *middle*, and A_k the *end*. Let us combine the starts and middles of all the triangles into a set D and let E be the set of ends. Now consider the first p points $A_1, \ldots, A_p$, where m is any integer in the range $1 \leq p \leq 3n$. Among these points there must be at least twice as many in D as there are in E (for if the end of a triangle is one of the points $A_1, \ldots, A_p$, then the start and middle of that triangle are also among the points $A_1, \ldots, A_p$). Conversely, suppose the points $A_1, \ldots, A_{3n}$ are divided into a set D of $2n$ points and a set E of n points in such a way that for any $p \leq 3n$ there are at least twice as many points of D among $A_1, \ldots, A_p$ as there are points of E. We will show that there is then one and only one way of drawing n non-intersecting triangles whose starts and middles are in D and whose ends are in E. Let k be the smallest integer such that A_k is in E (we have $k \geq 3$ since A_1 and A_2 must be in D to satisfy the hypothesis). Then A_{k-2} and A_{k-1} are the start and middle of the triangle whose end is A_k (for otherwise we would get two intersecting triangles). Now delete A_{k-2}, A_{k-1}, A_k and apply the same reasoning to the remaining points. In this way we can successively determine all the triangles, and our assertion is proved.

From this it follows that G_n, the number of ways of grouping the points $A_1, \ldots, A_{3n}$ into n non-intersecting triangles, is equal to the number of ways of dividing the points $A_1, \ldots, A_{3n}$ into two sets D and E (of $2n$ and n points respectively) so that there are at least twice as many points of D among $A_1, \ldots, A_p$ as there are points of E. If we think of the points of D as customers with one-dollar bills and the points of E as customers with three-dollar bills, we see that G_n is equal to the number of favorable outcomes in problem 83c with the n of that problem replaced by $2n$ and the m of that problem replaced by n. Thus

$$G_n = \frac{2n - 2n + 1}{2n + 1} \binom{2n + n}{n} = \frac{1}{2n + 1} \binom{3n}{n}.$$

We can also write G_n in any of the following forms:

$$G_n = \frac{(3n)!}{n!\,(2n + 1)!} = \frac{1}{3n + 1} \frac{(3n + 1)!}{n!\,(2n + 1)!}$$

$$= \frac{1}{3n + 1} \binom{3n + 1}{n} = \frac{1}{n} \binom{3n}{n - 1}$$

$$= \frac{3}{2n + 1} \binom{3n - 1}{n - 1} = \frac{3}{n - 1} \binom{3n - 1}{n - 2}.$$

84c. Denote by S_n the number of different ways of decomposing a convex $2n$-gon into quadrilaterals by means of diagonals which do not intersect within the $2n$-gon.[9] It will be shown below that $S_{n+1} = G_n$; by virtue of part b, we then see that $S_n = \binom{3n - 3}{n - 1} \Big/ (2n - 1)$.

---

[9] Following the solution to problem 52 it is easy to show that such a decomposition will involve $n - 1$ quadrilaterals and that $n - 2$ diagonals will be used in the decomposition. It follows from this that it is impossible to decompose a $(2n + 1)$gon into quadrilaterals by means of diagonals which do not intersect within it.

The proof that $S_{n+1} = G_n$ is carried out in the same way as the proof that $T_{n+2} = F_n$ (see the solution to problem 54). First of all we derive a relation which allows us to determine G_n from the values of G_1, G_2, G_3, ..., G_{n-1} then we show that the numbers $Q_n = S_{n+1}$ satisfy the same relation, and that $Q_1 = G_1$, after which the equality $S_{n+1} = G_n$ follows immediately.

Let us turn to the derivation of the relation connecting G_n with the numbers G_1, G_2, G_3, ..., G_{n-1}. Denote the $3n$ points treated in problem 84b in counterclockwise order around the circle by A_1, A_2, A_3, ..., A_{3n}. Suppose we have an admissible division of A_1, A_2, $\cdots A_{3n}$ into triangles; i.e., one in which no two triangles intersect. Consider whichever one of the

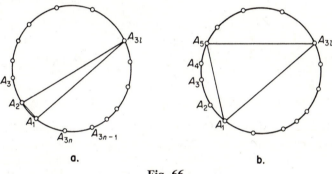

a.   b.

Fig. 66

n inscribed triangles with vertices at these points has A_1 as a vertex. It is clear that the second vertex A_k of this triangle (that is, second with respect to the order A_1, A_2, ..., A_{3n}) must be one of the points A_2, A_5, A_8, ..., A_{3k-1}, ..., A_{3n-1}. For otherwise the number of vertices within the arc cut off by A_1, A_k would not be a multiple of 3. Hence the triangles with vertices on that arc could not lie entirely on the same side of A_1, A_k, and so one of them would have to cross it. Suppose for example that the second vertex is the point A_2; then the third vertex must be one of the points A_3, A_6, A_9, ..., A_{3l}, ..., A_{3n}. If the third vertex is the point A_{3l} (fig. 66a), then the side $A_2 A_{3l}$ of the triangle in question will cut off an arc inside of which are located $3(l - 1)$ of our points (the points A_3, A_4, A_5, ..., A_{3l-1}) and the side $A_{3l} A_1$ will cut off an arc inside of which are located $3(n - l)$ of the points (the points A_{3l+1}, A_{3l+2}, ..., A_{3n}). We obtain all admissible divisions involving the triangle $A_1 A_2 A_{3l}$ by combining each of the G_{l+1} admissible divisions of A_3, A_4, ..., A_{3l+1} with each of the G_{n-l} admissible divisions of A_{3l+1}, A_{3l+2}, ..., A_{3n}. Consequently, the total number of admissible divisions involving the triangle $A_1 A_2 A_3$ is $G_{l-1} G_{n-l}$. (Here we are making the convention that $G_0 = 1$.)

Inasmuch as l can take any of the values $1, 2, 3, \ldots, n$, the total number of admissible divisions in which A_1 and A_2 belong to the same triple is equal to

$$G_{n-1} + G_1 G_{n-2} + G_2 G_{n-3} + \cdots + G_{n-2} G_1 + G_{n-1}$$

(the first and last terms correspond to the cases $l = 1$ and $l = n$; these terms are obviously equal to G_{n-1}). If the second vertex of the triangle containing A_1 is A_5, then the third vertex can be any of the points A_6, A_9, $\ldots, A_{3l}, \ldots, A_{3n}$; we obtain all admissible ways of dividing the $3n$ points into triples, one of which is (A_1, A_5, A_{3l}) (fig. 66b), by combining the $G_1 = 1$ ways of dividing the 3 points A_2, A_3, A_4 cut off by the side $A_1 A_5$ into admissible triples with the G_{l-2} ways of dividing the $3(l - 2)$ points $A_6, A_7, \ldots, A_{3l-1}$ cut off by the side $A_5 A_{3l}$ into admissible triples and the G_{n-l} ways of dividing the $3(n - l)$ points $A_{3l+1}, A_{3l+2}, \ldots, A_{3n}$ cut off by the side $A_{3l} A_1$ into admissible triples; therefore the total number of ways of dividing the $3n$ points into admissible triples one of which is (A_1, A_5, A_{3l}) is equal to $G_1 G_{l-2} G_{n-l}$. Letting l take successively the values $2, 3, 4, \ldots, n$, we obtain the sum

$$G_1(G_{n-2} + G_1 G_{n-3} + G_2 G_{n-4} + \cdots + G_{n-3} G_1 + G_{n-2})$$

for the total number of ways of dividing the points into admissible triples in such a way that the second vertex of the triangle containing A_1 is A_5. Continuing to argue in this manner, we find that the number of ways of dividing the $3n$ points into admissible triples such that the second vertex of the triangle containing A_1 is A_8 equals

$$G_2(G_{n-3} + G_1 G_{n-4} + \cdots + G_{n-4} G_1 + G_{n-3}),$$

etc. The number of ways of dividing the $3n$ points into admissible triples in such a way that the second vertex of the triangle containing A_1 is A_{3n-4} equals $G_{n-2}(G_1 + G_1)$, and finally, the number in which this vertex is A_{3n-1} equals G_{n-1}. Therefore we obtain the following general formula for G_n:[10]

$$
\begin{aligned}
G_n = {} & G_{n-1} + G_1 G_{n-2} + G_2 G_{n-3} + \cdots + G_{n-2} G_1 + G_{n-1} \\
& + G_1(G_{n-2} + G_1 G_{n-3} + \cdots + G_{n-3} G_1 + G_{n-2}) \\
& + G_2(G_{n-3} + G_1 G_{n-4} + \cdots + G_{n-4} G_1 + G_{n-3}) \\
& + \quad \cdot \qquad \cdot \qquad \cdot \qquad \cdot \qquad \cdot \qquad \cdot \qquad \cdot \\
& + G_{n-3}(G_2 + G_1 G_1 + G_2) + G_{n-2}(G_1 + G_1) + G_{n-1}.
\end{aligned}
$$

[10] By putting $G_0 = 1$ and using the summation sign, this formula can be abbreviated as follows:

$$G_n = \sum_{i=0}^{n-1} \sum_{j=0}^{n-i-1} G_i G_j G_{n-i-j-1} = \sum_{i+j+k=n-1} G_i G_j G_k.$$

This is the relation we wanted to obtain. Using it and taking into account the fact that $G_1 = 1$ (G_1 is the number of ways of dividing three points into triples, which can obviously be done in only one way), we can successively compute all the values of G_n; in particular, it follows from this that

$$G_2 = G_1 + G_1 + G_1 = 3,$$
$$G_3 = G_2 + G_1 G_1 + G_2 + G_1(G_1 + G_1) + G_2$$
$$= 3 + 1 \cdot 1 + 3 + 1 \cdot (1 + 1) + 3 = 12,$$
$$G_4 = G_3 + G_1 G_2 + G_2 G_1 + G_3 +$$
$$+ G_1(G_2 + G_1 G_1 + G_2) + G_2(G_1 + G_1) + G_3$$
$$= 12 + 1 \cdot 3 + 3 \cdot 1 + 12 + 1 \cdot (3 + 1 \cdot 1 + 3) +$$
$$+ 3 \cdot 1(1 + 1) + 12 = 55,$$

etc. (All these values can also be obtained from the formula for G_n which was derived in the solution to problem 84b.)

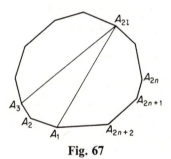

Fig. 67

Let us now derive a similar relation connecting S_{n+1} with S_n, S_{n-1}, $S_{n-2}, \ldots, S_2$. Let $A_1 A_2 \cdots A_{2n} A_{2n+1} A_{2n+2}$ be a convex $(2n + 2)$gon (fig. 67); for any decomposition of the $(2n + 2)$gon into quadrilaterals, consider the quadrilateral which contains the side $A_1 A_2$. The third vertex of this quadrilateral (that is, third with respect to the order A_1, A_2, A_3, $\ldots$, A_{2n+2}) must be one of the points A_3, A_5, A_7, $\ldots$, A_{2n+1} (an even-numbered vertex of the $(2n + 2)$gon could not be the third vertex of this quadrilateral, since otherwise the diagonal joining the second and third vertices would cut off from the $(2n + 2)$gon a polygon with an odd number of sides, and such a polygon cannot be divided into quadrilaterals by nonintersecting diagonals; see footnote on page 191). If this third vertex is A_3, then the fourth vertex must be one of the points A_4, A_6, A_8, $\ldots$, A_{2n+2}; the total number of decompositions of our $(2n + 2)$gon in which the quadrilateral $A_1 A_2 A_3 A_{2l}$ occurs is $S_{l-1} S_{n-l+2}$, since the sides $A_3 A_{2l}$ and $A_1 A_{2l}$, respectively, cut off a $2(l - 1)$gon and a $2(n - l + 2)$gon from the $(2n + 2)$gon (see fig. 67). It follows from this that the total number of decompositions in which the third vertex of the quadrilateral containing $A_1 A_2$ is A_3 equals

$$S_n + S_2 S_{n-1} + S_3 S_{n-2} + \cdots + S_{n-1} S_2 + S_n.$$

It can be proved in the same way that the total number of decompositions in which the third vertex is A_5 equals

$$S_2 (S_{n-1} + S_2 S_{n-2} + \cdots + S_{n-2} S_2 + S_{n-1}),$$

the total number of decompositions in which this vertex is A_7 equals

$$S_3 (S_{n-2} + S_2 S_{n-3} + \cdots + S_{n-3} S_2 + S_{n-2}),$$

etc.; the number of decompositions in which this vertex is A_{2n-3} equals

$$S_{n-2} (S_3 + S_2 S_2 + S_3);$$

the number of decompositions in which this vertex is A_{2n-1} equals

$$S_{n-1} (S_2 + S_2),$$

and finally, the number of decompositions in which this vertex is A_{2n+1} equals S_n.

We obtain from this the following formula for S_{n+1}:[11]

$$\begin{aligned}
S_{n+1} = {} & S_n + S_2 S_{n-1} + S_3 S_{n-2} + \cdots + S_{n-1} S_2 + S_n \\
& + S_2 (S_{n-1} + S_2 S_{n-2} + \cdots + S_{n-2} S_2 + S_{n-1}) \\
& + S_3 (S_{n-2} + S_2 S_{n-3} + \cdots + S_{n-2}) \\
& + \cdot \quad \cdot \quad \cdot \quad \cdot \quad \cdot \quad \cdot \\
& + S_{n-2} (S_3 + S_2 S_2 + S_3) + S_{n-1} (S_2 + S_2) + S_n.
\end{aligned}$$

This is the formula we required. Using this formula and taking into consideration that $S_2 = 1$, we can compute S successively for all values of n; in particular, for $n = 3, 4, 5$ it follows from this that

$$\begin{aligned}
S_3 = {} & S_2 + S_2 + S_2 = 3, \\
S_4 = {} & S_3 + S_2 S_2 + S_3 + S_2 (S_2 + S_2) + S_3 \\
= {} & 3 + 1 \cdot 1 + 3 + 1 \cdot (1 + 1) + 3 = 12, \\
S_5 = {} & S_4 + S_2 S_3 + S_3 S_2 + S_4 + S_2 (S_3 + S_2 S_2 + S_3) \\
& + S_3 (S_2 + S_2) + S_4 \\
= {} & 12 + 3 + 3 + 12 + (3 + 1 + 3) + 3 \cdot (1 + 1) + 12 = 55.
\end{aligned}$$

[11] By putting $S_1 = 1$ and using the summation sign, we can abbreviate this formula as follows:

$$S_{n+1} = \sum_{i=1}^{n} \sum_{j=1}^{n-i+1} S_i S_j S_{n-i-j+2} = \sum_{i+j+k=n+2} S_i S_j S_k.$$

Note that if we denote S_{n+1} by Q_n, the relation just derived assumes the following form:

$$Q_n = Q_{n-1} + Q_1 Q_{n-2} + Q_2 Q_{n-3} + \cdots + Q_{n-2} Q_1 + Q_{n-1}$$
$$+ Q_1(Q_{n-2} + Q_1 Q_{n-3} + \cdots + Q_{n-3} Q_1 + Q_{n-2})$$
$$+ Q_2(Q_{n-3} + Q_1 Q_{n-4} + \cdots + Q_{n-3})$$
$$+ \cdot \quad \cdot \quad \cdot \quad \cdot \quad \cdot \quad \cdot \quad \cdot \quad \cdot$$
$$+ Q_{n-3}(Q_2 + Q_1 Q_1 + Q_2) + Q_{n-2}(Q_1 + Q_1) + Q_{n-1}.$$

This relation is the same as the relation which was obtained above for the values of G_n (see p. 188). Since, furthermore, $Q_1 = S_2 = 1 = G_1$, the successive computation of the values of Q_n according to our formula for $n = 2, 3, 4, \ldots$ will give exactly the same results as are obtained in the successive computation of the values of G_n; in other words,

$$Q_n = G_n,$$

that is,

$$S_n = Q_{n-1} = G_{n-1}.$$

This is the answer to our problem.

Remark. Problems 54 and 84b are special cases of the following more general problem.

kn points on the circumference of a circle are given. In how many ways can these points be divided into n groups of k points each in such a way that the sides of the n inscribed k-gons determined by these groups of k points do not intersect each other?

This general problem can be solved in exactly the same way that problem 84b was solved; the solution differs only in that it is based not on problem 84c but on the generalization mentioned in the remark following the solution to problem 84c. The required number of ways is

$$\frac{1}{(k-1)n+1}\binom{kn}{n} = \frac{1}{kn+1}\binom{kn+1}{n} = \frac{(kn)!}{n!\,[(k-1)n+1]!}$$

Similarly, Euler's problem and problem 84c are special cases of the following problem:

In how many different ways can a convex $[(k-2)n+2]$gon[12] be decomposed into k-gons by means of diagonals which do not intersect inside the $[(k-2)n+2]$gon?

---

[12] It is not hard to see that for $m \neq (k-2)n + 2$ there is no way of decomposing a convex m-gon into k-gons with diagonals which do not intersect within the m-gon. This follows, for example, from the fact that if an m-gon is decomposed into n k-gons, the sum of the angles of the m-gon must equal the sum of the angles of the k-gons, that is, n times the sum of the angles of a k-gon, whence $(m-2)180° = n \cdot (k-2)180°$, $m = (k-2)n + 2$.

As in the solution to problem 53b and 84c, this problem can be reduced without difficulty to the solution of the above problem. Using the answer given above to the generalized version of problem 84b, we find that the required number of ways is

$$\frac{1}{kn + 1} \binom{(k + 1)n}{n} = \frac{[(k + 1)n]!}{n! \, (kn + 1)!}.$$

This result is a generalization of the results found in the solutions to problems 53b and 84c.

85. We proceed at once to the solution of part b, since part a is the special case $k = 0$. If we denote a winning card by W and a losing card by L, then the experiment under consideration consists of choosing at random one of the $\binom{m + n}{m}$ sequences which can be formed with m W's and n L's. Let S be the set of all such sequences. A favorable outcome is a sequence σ with the property that at exactly k places in it, there is a negative amount of money in the bank. Let S_k be the set of all such outcomes. Then clearly $S = S_0 \cup S_1 \cup S_2 \cup \cdots \cup S_{m+n-1}$. We are going to show that $\# S_k$ is independent of k, i.e., $\#S_0 = \#S_1 = \cdots = \#S_{m+n-1}$. It then follows at once that the desired probability $p_k = \dfrac{\#S_k}{\#S} = \dfrac{1}{m + n}$.

To prove that $\#S_k$ is independent of k, we introduce an equivalence relation (see footnote, p. 121) into the set S as follows. Two sequences σ_1 and σ_2, are called equivalent if they differ only by "cyclic permutation"; i.e., if σ_2 can be gotten by moving a block of letters from the beginning of σ, to the end. For example, the sequences $(WLWWL)$, $(LWWLW)$, $(WWLWL)$, $(WLWLW)$, and $(LWLWW)$ are cyclic permutations of each other. It is easily seen that this is indeed an equivalence relation, and therefore S is partitioned into equivalence classes as explained on p. 121. Thus in the case $m = 3$, $n = 2$, the five sequences given above form one equivalence class. We will show that when m and n are relatively prime, each equivalence class consists of $m + n$ sequences, one of which is in S_0, one in S_1, and one in S_2, $\cdots$, and one in S_{m+n-1}. In the above example $(WLWWL)$ is in S_1, $(LWWLW)$ in S_3, $(WWLWL)$ in S_0, $(WLWLW)$ in S_2, and $(LWLWW)$ in S_4.

To make the idea of the proof intuitively clear, it is convenient to represent our sequences geometrically in much the same way as in problem 83. Starting at the origin A_0 we construct a path $P = A_0A_1A_2 \cdots A_{m+n}$ by moving one unit horizontally whenever a winning card is turned over, and one unit vertically whenever a losing card is turned over. For example, when $m = 3$, $n = 2$, figure 68a shows the path corresponding to the sequence $\sigma = (WLWWL)$.

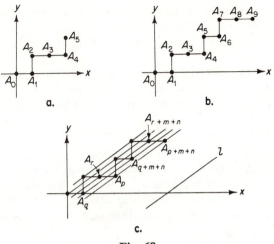

Fig. 68

The possible outcomes of the experiment are thus represented by paths from $(0,0)$ to (m,n), composed of m horizontal and n vertical unit segments. Now suppose σ is in S_k; we want to see what geometric condition the corresponding path P will then satisfy. Let $A_{i+j} = (i,j)$ be any point of P. Then the amount of money in the bank at the $(i + j)$'th stage of the game is $-ai + bj$, since i people have withdrawn a dollars each, and j people have paid in b dollars each. Thus the condition for the bank to contain a negative amount of money at this stage is that $bj < ai$, or in other words

$$\frac{j}{i} < \frac{a}{b} = \frac{n}{m}$$

(remembering that $am = bn$ by hypothesis). Geometrically, this says that the slope of A_0A_{i+j} is less than that of A_0A_{m+n}, so that the point A_{i+j} is below the line A_0A_{m+n}. Thus the paths P corresponding to sequences σ in S_k are those having exactly k vertices below the line A_0A_{m+n}. (By the "vertices" of P we mean the points $A_0, A_1, \ldots, A_{m+n}$; these need not be corners.)

Now let σ_0 be a given sequence, and let $P_0 = A_0A_1 \cdots A_{m+n}$ be the corresponding path.

Extend P_0 to a path $P' = A_0A_1 \cdots A_{m+n} A_{m+n+1} \cdots A_{2m+2n-1}$ of length $2m + 2n - 1$ by drawing a path congruent to $A_0A_1 \cdots A_{m+n-1}$ with its initial point at A_{m+n}. (In figure 68b this procedure is illustrated for the path of figure 68a.) The paths $P_1 = A_1A_2 \cdots A_{m+n+1}$, $P_2 = A_2A_3 \cdots A_{m+n+2}$, $\cdots$, $P_{m+n-1} = A_{m+n-1} A_{m+n} \cdots A_{2m+2n-1}$ correspond to the cyclic permutations of the sequence σ_0.

Therefore, what we are trying to prove can be interpreted geometrically as follows: For each k $(0 \leq k \leq m + n - 1)$, we can find one of the paths P_t with exactly k vertices below the chord $A_t A_{t+m+n}$ joining its endpoints.

To prove this, we first show that for any $h < m + n$, the line joining $A_h = (i,j)$ and $A_{h+m+n} = (i + m, j + n)$ contains no further vertices of P'. For suppose there were such a vertex (r,s). Then by similar triangles, we would have $\dfrac{s - j}{r - i} = \dfrac{n}{m}$. Since m and n are relatively prime, this implies that there is an integer q such that $s - j = qn$, $r - i = qm$. But since P' has only $2n - 1$ vertical segments, we have $s - j < 2n$, and since $s \geqslant 0$, $j < n$, we have $s - j > -n$. Thus the only possible values for q are $q = 0$ and $q = 1$. If $q = 0$, then $(r,s) = (i,j)$, while if $q = 1$, then $(r,s) = (i + m, j + n)$. Thus (r,s) cannot be distinct from A_h and A_{h+m+n}.

Now draw a line l parallel to $A_0 A_{m+n}$ and situated to the right of all the points of P' (fig. 68c). Move l to the left until it first touches P'. Since l is parallel to $A_0 A_{m+n}$, it will at this moment pass through two vertices A_p and A_{p+m+n}, and as shown above, will contain no other vertices of P'. The path P_p corresponds to an outcome in S_0, since there are no vertices below $A_p A_{p+m+n}$.

Let l continue moving to the left until it strikes another pair of vertices A_q, A_{q+m+n}. This time the path P_q corresponds to an outcome in S_1, for P_q contains one and only one of the vertices A_p, A_{p+m+n}. Indeed, if A_p is not a vertex of P_q, then $p < q$, so that $q < p + m + n < q + m + n$, showing A_{p+m+n} is a vertex of P_q. This reasoning is reversible.

As l continues moving to the left it strikes another pair of vertices A_r, A_{r+m+n}. The path P_r corresponds to an outcome in S_2, for (reasoning as above) it contains one of the vertices A_p, A_{p+m+n} and one of the vertices A_q, A_{q+m+n}. Proceeding in this way we get for each k, $0 \leq k \leq m + n - 1$, a path P_t with exactly t vertices below the chord $A_t A_{t+m+n}$.

This shows that each equivalence class of sequences has $m + n$ members, one in S_0, one in S_1, ..., and one in S_{m+n-1}. Hence $\#S_0 = \#S_1 = \cdots = \#S_{m+n-1}$, which is what we wanted to prove.

VII. EXPERIMENTS WITH INFINITELY MANY OUTCOMES

86. Of any six consecutive integers, one is divisible by 6, one gives a remainder of 1 upon division by 6, one gives a remainder of 2, one gives a remainder of 3, one a remainder of 4, and one a remainder of 5. Therefore, of any six consecutive integers, exactly two are relatively prime to 6 (those which give remainders of 1 and 5 upon division by 6). Let N be any positive integer and represent it in the form $N = 6Q + R$. Since the integers from 1 to Q split into Q groups of six consecutive integers each, exactly $2Q$ integers from 1 to $6Q$ are relatively prime to 6, and of the next R integers, at most two can be relatively prime to 6 (since R is at most 5). Thus a total of $2Q + r$ integers from 1 to N are relatively prime to 6, where r is 0, 1, or 2. Consequently, the probability that a number selected at random from the positive integers up to N is relatively prime to 6 equals $P(N) = (2Q + r)/N = (2Q + r)/(6Q + R)$. As $N \to \infty$, this expression approaches the limit $2/6 = 1/3$. Therefore the probability that a positive integer selected at random is relatively prime to 6 equals $1/3$.

The probability that at least one of two integers selected at random from the integers from 1 to N will be relatively prime to 6 is equal to the probability that the first integer drawn is prime to 6, plus the probability that the second integer drawn is prime to 6, minus the probability that both are prime to 6; that is,

$$2P(N) - P(N)^2 = P(N)(2 - P(N)) = \frac{(2Q + r)(10Q + 2R - r)}{(6Q + R)^2}.$$

As $N \to \infty$, this expression approaches the limit $2 \cdot 10/6^2 = 5/9$, which is thus the probability that at least one of two numbers drawn at random from all positive integers will be relatively prime to 6 (compare with the solution to problem 73a).

87a. Thus n^2 ends with a 1 if and only if n ends with a 1 or a 9. Of any ten consecutive integers, exactly two have this property, and so the required probability is $2/10 = 1/5$. We will show that the cube of an integer n ends in 11 if and only if n ends in 71. To see this write n in the form $n = 100q + r$, where $0 \leqq r \leqq 99$. Then

$$n^3 = 1,000,000q^3 + 30,000q^2r + 300qr^2 + r^3.$$

The first three terms of this expression all end in 00, and therefore the last two digits of n^3 are the same as those of r^3.

If r ends in a 1, so does r^3; but if r ends in a 2, 3, 4, 5, 6,7 , 8, 9, or 0, then r^3 ends in 8, 7, 4, 5, 6, 3, 2, 9, or 0 respectively. Thus the only values of r whose cubes could possibly end in 11 are 1, 11, 21, 31, 41, 51, 61, 71, 81, or 91. The cubes of these numbers end in 01, 31, 61, 91, 21, 51, 81, 11, 41, and 71 respectively. Thus n^3 ends in 11 if and only if n ends in 71. The probability of this is $1/100$, since one integer in every hundred ends with 71.

87b. It is obvious that the last digit of the 10th power of an integer n depends only on the last digit of n. Further, it is easy to verify that 4^{10} and 6^{10} end in the digit 6 and the 10th powers of the remaining one-digit numbers end in figures other than 6 (2^{10} and 8^{10} end in 4's, $0^{10} = 0$, and the odd numbers have odd 10th powers). Hence, of any 10 consecutive positive integers, exactly two have 10th powers which end in the digit 6. Consequently, the probability that the 10th power of an integer selected at random will end in a 6 is $2/10 = 0.2$.

Since the 10th power of any even number other than a multiple of 10 ends in a 4 or a 6, and since $4^2 = 16$ and $6^2 = 36$, the 20th power of any such number ends in a 6. All other positive integers have 20th powers which end in digits other than 6 (they end in either a 0 or an odd digit). It follows from this that the probability that n^{20} ends in a 6 is $4/10 = 0.4$.

Remark. It can similarly be proved that the probability that the 20th power of an integer ends in 76 is 0.4, and that the probability that the 200th power of a positive integer ends in 376 is also 0.4. In this connection, see the solution to problem 34 of the book *Izbrannye zadachi i teoremy elementarnoi matematiki*, vol. 1, by D. O. Shklyarskii, N. N. Tschentsov, and I. M. Yaglom.

88. $\dbinom{n}{7} = \dfrac{n(n-1)(n-2)(n-3)(n-4)(n-5)(n-6)}{1 \cdot 2 \cdot 3 \cdot 4 \cdot 5 \cdot 6 \cdot 7}$; hence the

probability that $\dbinom{n}{7}$ is divisible by 7 equals the probability that the product $n(n-1)(n-2)(n-3)(n-4)(n-5)(n-6)$ is divisible by 49. But the latter can occur only if one of the seven factors is divisible by 49, that is, if n gives a remainder of 0, 1, 2, 3, 4, 5, or 6 upon division by 49. Of any 49 consecutive integers, exactly seven satisfy this condition. It follows from this that the required probability is $7/49 = 1/7$.

The probability that $\dbinom{n}{7}$ is divisible by 12 is the same as the probability that the product $n(n-1)(n-2)(n-3)(n-4)(n-5)(n-6)$ in the numerator is divisible by $12 \cdot 2 \cdot 3 \cdot 4 \cdot 6 = 64 \cdot 27$. Of any seven consecutive integers, at least two are divisible by 3; the product thus is divisible by 27 if and only if another factor of 3 appears in it, that is, if one of the factors is divisible by 9 or three of the seven factors are divisible by 3. But the latter case is included in the former, since of any three consecutive multiples of 3, one has to be a multiple of 9. Consequently, the product $n(n-1)(n-2)(n-3)(n-4)(n-5)(n-6)$ is divisible by 27 if and only if n has the form $9k + r$, where $r = 0, 1, 2, 3, 4, 5$, or 6.

Furthermore, of our seven consecutive integers, either $n-1$, $n-3$, and $n-5$ are all divisible by 2 or else n, $n-2$, $n-4$, and $n-6$ are. In the first case, of the three consecutive even numbers $n-1$, $n-3$, and $n-5$, either $n-3$ is divisible by 4 or both $n-1$ and $n-5$ are divisible by 4. If $n-1$ and $n-5$ are divisible by 4, then one of these two numbers has to be a multiple of 8; hence, if $n-1$ is divisible by 4 (that is, n has the form $4l + 1$), then our product will be divisible by $2 \cdot 4 \cdot 8 = 64$. If $n-3$ is divisible by 4, then in order that our product be divisible by 64, it is necessary that $n - 3$ be divisible by 16, that is, that n have the form $16l + 3$. Finally, if n, $n - 2$, $n - 4$, and $n - 6$ are divisible by 2 (that is, if $n = 2l$ is even), then two of these four consecutive even numbers will have to be divisible by 4. This means that $n(n - 2)(n - 4)(n - 6)$ is divisible by $2 \cdot 2 \cdot 4 \cdot 4 = 64$ (it is easy to see that in this case it is even divisible by 128).

Thus, $n(n - 1)(n - 2)(n - 3)(n - 4)(n - 5)(n - 6)$ is divisible by 64 if and only if n has the form $4l + 1$ or $16l + 3$ or $2l$, or in other words, if and only if n has the form $16l + s$, where $s = 0, 1, 2, 3, 4, 5, 6, 8, 9, 10, 12, 13,$ or 14.

Thus, $\binom{n}{7}$ is divisible by 12 if and only if n has the form $9k + r$, where r can take any of the seven values listed above, and simultaneously has the form $16l + s$, where s can take any of the 13 values listed above.

Accordingly, the values of n for which $\binom{n}{7}$ are divisible by 12 are those which gave any of $7 \cdot 13 = 91$ remainders upon division by $9 \cdot 16 = 144$. For example, if n gives a remainder of 8 upon division by 16 and a remainder of 5 upon division by 9, then n has to have the form $144m + 104$: of the numbers of the form $144m + t$, where $t = 8, 24, 40, 56, 72, 88, 104, 120, 136$, only the numbers of the form $144m + 104$ give a remainder of 5 upon division by 9. This is a special case of the *Chinese remainder theorem*: if d_1 and d_2 are relatively prime integers and $0 \leq r_1 < d_1$ and $0 \leq r_2 < d_2$, then there is exactly one nonnegative integer less than $d_1 d_2$ which gives a remainder of r_1 upon division by d_1 and a remainder of r_2 upon division by d_2.[1]

It follows from this that the required probability is $91/144 \approx 0.63$.

89. The first five powers of 2 are 2, 4, 8, 16, 32. Since $2^5 = 32$ has 2 for its last digit, $2^6 = 2 \cdot 2^5$ has 4 for its last digit, $2^7 = 2 \cdot 2^6$ has 8 for its last digit, $2^8 = 2 \cdot 2^7$ has 6 for its last digit, etc.; thus, the final digit of the successive powers of 2 keeps repeating cyclically in groups of four: 2, 4, 8, 6, 2, 4, 8, 6, ... Consequently, of any four consecutive powers of 2, exactly one ends in a 2; this means that the probability that 2^n ends in a 2 is $1/4 = 0.25$.

Now let us compute the last two digits of the successive powers of 2:

$$02, 04, 08, 16, 32, 64, 28, 56, 12, 24, 48, 96,$$
$$92, 84, 68, 36, 72, 44, 88, 76, 52, 04, 08, \ldots$$

(this is not hard to do, since it suffices each time to double the last two-digit number of the sequence and then, if the result is a three-digit number, to discard the hundreds-digit). Thus we see that 2^{22} is the first power of 2 after $2^2 = 4$ which ends in the figures 04. From there on the sequence

[1] See, for example, Hardy and Wright, *op. cit.*, p. 95.

keeps repeating cyclically: 2^{23} ends in the figures 08, 2^{24} ends in the figures 16, etc. Thus the 2nd through 21st members of this sequence are repeated as the 22nd through 41st members, then as the 42nd through 61st members, etc.; that is, they repeat periodically in groups of 20. Furthermore, of the 20 powers of 2 from $2^2 = 04$ to 2^{21}, only 2^9 ends in the digits 12. Thus, of any 20 consecutive powers of 2, exactly one ends in the digits 12. Consequently, the probability that 2^n ends in the digits 12 is $1/20 = 0.05$

Remark. It can be proved that the last k digits of the number 2^n are repeated in groups of $4 \cdot 5^{k-1}$, starting with the number 2^k (see Shklyarskii, Tschentsov, and Yaglom, *Izbrannye zadachi i teoremy elementarnoi matematiki*, vol. 1, problem 243).

90. Let $q(N)$ be the number of powers of 2 from 1 to 2^N which begin with the digit 1. We have to prove that $\lim_{N \to \infty} q(N)/N$ exists, and determine its value.

Note that there cannot be two different powers of 2 which have the same number of digits and which both begin with a 1. If two distinct powers of 2 are given, the larger must be at least twice the smaller; hence if both have the same number of digits, then the first digit of the larger must be at least twice the first digit of the smaller, which means that they cannot both begin with the digit 1. Furthermore, the *smallest* of all powers of 2 which have a given number of digits must begin with a 1: otherwise we could divide the number by 2 and obtain a smaller power of 2 which had the same number of digits (for example, the smallest 2-digit power of 2 is 16, the smallest 3-digit power of 2 is 128, the smallest 4-digit power of 2 is 1024, etc.). It is essential to note further that there are powers of 2 having any given number of digits: for if 2^k is the *greatest* power of 2 which has p digits, then the next power 2^{k+1} will have $p + 1$ digits; consequently (by mathematical induction) for arbitrary p there will be p-digit powers of 2.

Now let 2^N be an n-digit number. Then among the first N powers of 2 there will be one n-digit number which starts with a 1, one $(n - 1)$-digit number which starts with a 1, one $(n - 2)$-digit number which starts with a 1, etc., and finally one 2-digit number which starts with a 1 (namely, 16). Therefore, $q(N) = n - 1$.

If n is the number of digits in the number 2^N, then $n - 1$ is the characteristic of the logarithm (to the base 10) of 2^N, that is, the integral part of the number $\log 2^N = N \log 2$. In other words, $N \log 2 = n - 1 + \alpha_N$, where $0 \leq \alpha_N < 1$, and $q(N) = n - 1 = N \log 2 - \alpha_N$. This means that

$$\lim_{N \to \infty} \frac{q(N)}{N} = \lim_{N \to \infty} \frac{N \log 2 - \alpha_N}{N} = \log 2 - \lim_{N \to \infty} \frac{\alpha_N}{N} = \log 2,$$

that is, the probability that a power of 2 begins with the digit 1 is log 2 ≈ 0.30103.

91a. The problem can be reformulated as follows: it is required to prove that for any positive integer M there exists an n such that 2^n begins with the sequence of digits which represents the number M in the decimal notation. This is equivalent to proving that for any positive integer M one can find two positive integers n and k such that

$$10^k \cdot M \leq 2^n < 10^k(M + 1). \tag{1}$$

For the numbers beginning with the digits of M are precisely those of the form $10^k M + r$, where $0 \leq r < 10^k$. Taking logarithms of both sides of inequality (1) we obtain the equivalent inequality

$$\log M + k \leq n \log 2 < \log (M + 1) + k. \tag{2}$$

We will now prove the existence of positive integers n and k which satisfy (2).

Mark off on the real line the intervals[2] [$\log M + k$, $\log (M + 1) + k$), where k assumes all positive integral values. All these intervals have the same length, namely $\log (M + 1) - \log M = \log (M + 1)/M = \log (1 + 1/M)$ and all are obtained from the interval [$\log M$, $\log (M + 1)$) by translating it through positive integral distances 1, 2, 3, ... The numbers $\log 2$, $2 \log 2$, $3 \log 2$, ..., $n \log 2$, ... form an arithmetic progression; we have to prove that at least one of them lies in one of the intervals we have marked off.

It is convenient to imagine the real line as being wound around a circle of radius $1/2\pi$ (that is, of circumference 1). Then points differing by a positive integer will coincide and, in particular, all the intervals constructed above will coincide (fig. 69). As regards the points $\log 2$, $2 \log 2$, $3 \log 2$, ..., no two of them will coincide: if the points $p \log 2$ and $q \log 2$ were to coincide, the difference $p \log 2 - q \log 2$ would be an integer r; then we would have $\log 2 = r/(p - q)$, where r, p, and q are integers, which would contradict the fact that $\log 2$ is an irrational number. Consequently, the values of $\log 2$, $2 \log 2$, $3 \log 2$, ... form an infinite sequence of distinct points A_1, A_2, A_3, ... (see fig. 69, where the first 15 points A_1, A_2, A_3, ..., A_{15} are illustrated). We have to prove that there are points of the sequence A_1, A_2, A_3, ... which fall in the interval I on the circle corresponding to the interval [$\log M$, $\log (M + 1)$).

Since there are infinitely many distinct points of our sequence, it is possible to find a pair of such points whose distance apart (that is, the length of the smaller arc of the circle which has these points as end points) is less than any given number. For if the distance between every two

---

[2] By [a, b), we mean the set of all real numbers x such that $a \leq x < b$.

points of the sequence were greater than some positive number c, then only a finite number of points could be accommodated on the circle. Let A_p and A_{p+q} be two points of our sequence whose distance apart is less than $a = \log(1 + 1/M)$, the length of the interval I. Note that the distance between the points A_p and A_{p+q} equals the distance between A_{p+q} and A_{p+2q}, between A_{p+2q} and A_{p+3q}, between A_{p+3q} and A_{p+4q}, etc. (This follows from the fact that $(p+q)\log 2 - p \log 2 = q \log 2 = (p+2q)\log 2 - (p+q)\log 2 = (p+3q)\log 2 - (p+2q)\log 2 = \cdots$.) Therefore, the points A_p, A_{p+q}, A_{p+2q}, A_{p+3q}, ... are the same distance apart on the circle. And since the distance between any two adjacent points of this subsequence is less than the length a of the interval I, at least one of any k

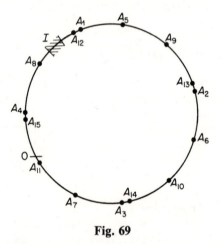

Fig. 69

consecutive points in this subsequence, where k is a positive integer such that $ka > 1$, will have to lie in I. (See fig. 69, where one can take, say, the points A_1, A_{12}, A_{23}, A_{34}, ... in the role of the sequence of points A_p, A_{p+q}, A_{p+2q}, ...; the first two of these points are represented in the figure.) This concludes the proof.

91b. Using the geometric interpretation introduced in part a, we can reformulate the problem as follows: what is the probability that a point selected at random from the sequence A_1, A_2, A_3, ... will lie in the interval I constructed above? We will prove that the desired probability is the length $a = \log(1 + 1/M)$ of I. More precisely, let N be a positive integer, and let $f(N)$ be the number of points of the sequence A_1, A_2, ..., A_N which lie in I. We shall prove that $\lim_{N \to \infty} f(N)/N = a$.

Given a positive number $\varepsilon < \frac{1}{2}$, we know from part a that there exist two points A_p and A_{p+q} whose (circular) distance d apart is $<\varepsilon$. For any i, the distance between A_i and A_{i+q} is also d, since the pair $(A_i \ A_{i+q})$

can be obtained from (A_p, A_{p+q}) by rotating the circle. Now let n be the (unique) integer such that $1/n > d \geq 1/(n + 1)$. The points A_i, A_{i+q}, $A_{i+2q}, \ldots, A_{i+(n-1)q}$ are located around the circle at a distance of d from one to the next. Therefore the arc $A_i A_{i+q} \ldots A_{i+(n-1)q}$ has length $(n - 1)d$, and since the circle has a circumference 1, the distance from $A_{i+(n-1)q}$ to A_i is $1 - (n - 1)d = e$. Since $d < 1/n$, we have $e > 1 - (n - 1)/n = 1/n$, and since $d \geq 1/(n + 1)$, we have $e \leq 1 - (n - 1)/(n + 1) = 2/(n + 1)$. Thus $d < e \leq 2d$. Now let m be the integer such that $m/n \leq a < (m + 1)/n$ (note that $m < n$ since $a < 1$). Then at most $m + 2$ and at least $m - 1$ of the points A_i, A_{i+q}, $A_{i+2q}, \ldots, A_{i+(n-1)q}$ lie in I. For if there are k of these points in I, they span an arc σ of length $(k - 1)d$ contained in I. Consequently

$$(k - 1)d < a < \frac{m + 1}{n}; \quad k - 1 < \frac{m + 1}{n} \frac{1}{d},$$

and since

$$\frac{1}{d} \leq n + 1, \quad k - 1 < \frac{m + 1}{n} (n + 1) = (m + 1) \left(1 + \frac{1}{n} \right)$$
$$= m + 1 + \frac{m + 1}{n}.$$

Since $m < n$, we have $(m + 1)/n \leq 1$, so that $k - 1 < m + 2$. Since $k - 1$ is an integer, this implies $k - 1 \leq m + 1$, or $k \leq m + 2$. On the other hand, I is certainly covered by the arc τ obtained from σ by adjoining an arc of length d to one of its ends and an arc of length e to its other end. Since τ has length $kd + e \leq (k + 2)d$, we have

$$(k + 2)d \geq a \geq \frac{m}{n}, \quad \text{or} \quad k + 2 \geq \frac{m}{n} \frac{1}{d}$$

Since $1/d > n$, it follows that $k + 2 > m$, whence $k \geq m - 1$.

Now consider the points $A_1, A_2, \ldots, A_{nq}$. These can be divided into q sets of the form $\{A_i, A_{i+q}, A_{i+2q}, \ldots, A_{i+(n-1)q}\}$ as follows:

$$S_1 = \{A_1, A_{1+q}, A_{1+2q}, \ldots, A_{1+(n-1)q}\}$$
$$S_2 = \{A_2, A_{2+q}, A_{2+2q}, \ldots, A_{2+(n-1)q}\}$$
$$\vdots$$
$$S_q = \{A_q, A_{2q}, A_{3q}, \ldots, A_{nq}\}$$

We have just shown that each set S_i contains at least $m - 1$ and at most $m + 2$ points of I. Therefore there are at least $q(m - 1)$ and at most $q(m + 2)$ points of I among $A_1, A_2, \ldots, A_{nq}$. The same reasoning can be applied to any nq consecutive points $A_{j+1}, A_{j+2}, \ldots, A_{j+nq}$.

For any positive integer N we can write $N = snq + t$, where $s = [N/nq]$ is the quotient obtained by dividing nq into N, and the remainder t satisfies $0 \leq t < nq$. We can then divide the points $A_1, A_2, \ldots, A_{snq}$ into s blocks of the form $A_{j+1}, A_{j+2}, \ldots, A_{j+nq}$ (where $j = 0, nq, 2nq, \ldots,$

$(s - 1)nq)$. Each of these s blocks contains at least $q(m - 1)$ and at most $q(m + 2)$ points of I. We do not know how many of the remaining t points $A_{snq+1}, A_{snq+2}, \ldots, A_N$ are in I, but in any case it follows that $sq(m - 1) \leqq f(N) \leqq sq(m + 2) + t$. Since $N = snq + t$, we have $sq = (N - t)/n$. Substituting this, we obtain

$$\frac{N - t}{n}(m - 1) \leqq f(N) \leqq \frac{N - t}{n}(m + 2) + t$$

or, dividing by N,

$$\left(1 - \frac{t}{N}\right)\frac{m - 1}{n} \leqq \frac{f(N)}{N} \leqq \left(1 - \frac{t}{N}\right)\frac{m + 2}{n} + \frac{t}{N}.$$

Since $m/n \leqq a < (m + 1)/n$, this implies that

$$\left(1 - \frac{t}{N}\right)\left(a - \frac{2}{n}\right) \leqq \frac{f(N)}{N} \leqq \left(1 - \frac{t}{N}\right)\left(a + \frac{2}{n}\right) + \frac{t}{N}.$$

As $N \to \infty$, the quantity t/N tends to 0, since t remains bounded ($t < nq$). Hence the left side tends to $a - 2/n$, and the right side to $a + 2/n$. It follows that for sufficiently large N,

$$a - \frac{3}{n} < \frac{f(N)}{N} < a + \frac{3}{n}, \quad \text{or} \quad \left|\frac{f(N)}{N} - a\right| < \frac{3}{n}.$$

Now for any positive integer n, $1/n \leqq 2/(n + 1)$. Hence

$$\left|\frac{f(N)}{N} - a\right| < \frac{6}{n + 1} \leqq 6d < 6\varepsilon.$$

Thus we have shown that for any $\varepsilon > 0$, the inequality

$$\left|\frac{f(N)}{N} - a\right| < 6\varepsilon$$

holds for all sufficiently large N. But this means that

$$\lim_{N \to \infty} \frac{f(N)}{N} = a.$$

Therefore, the probability that a power of 2 begins with the digits which represent the number M (in decimal notation) is $\log(1 + 1/M)$. In particular, if $M = 1$, then we obtain again the result of problem 88: the probability that 2^n begins with a 1 is $\log(1 + 1/1) = \log 2$.

Remark. It is curious to observe that for *any positive integer A* whose logarithm to the base 10 is irrational (that is, any positive integer other than a power of 10) this same expression gives the probability that a power A^n picked at random will begin with the digits which represent the number M. The proof of this does not differ essentially from the proof for the case of powers of 2.

92· The experiment which consists of choosing two integers a, b with $1 \leqq a, b \leqq N$ has N^2 equally likely possible outcomes (since there are N

possibilities for a and N for b). We have to compute the number $f(N)$ of favorable outcomes, i.e., outcomes in which a and b are relatively prime. To do this we will first compute the number $g(N)$ of unfavorable outcomes; then $f(N) = N^2 - g(N)$. Let $p_1, p_2, \ldots, p_m$ be the primes $\leq N$, arranged in increasing order, i.e. $2 = p_1 < p_2 < p_3 < \cdots < p_m \leq N$. If a and b are not relatively prime, then they are both divisible by some p_i, and conversely. Let A_i be the set of all ordered pairs a,b such that both a and b are divisible by p_i. Then $A_1 \cup A_2 \cup \cdots \cup A_m$ is the set of unfavorable outcomes, and hence $g(N) = \#(A_1 \cup A_2 \cup \cdots \cup A_m)$. We can compute this number by the principle of inclusion and exclusion, and for this purpose we now calculate the quantities $\#(A_i)$, $\#(A_i \cap A_j)$, $\#(A_i \cap A_j \cap A_k)$, etc. By definition $\#(A_i)$ is the number of ordered pairs a,b such that both a and b are divisible by p_i. The number of multiples of p_i between 1 and N is $[N/p_i]$; hence $\#(A_i) = [N/p_i]^2$. Next, observe that $A_i \cap A_j$ consists of ordered pairs a,b such that a and b are multiples of p_i and p_j, i.e., they are multiples of $p_i p_j$. Since there are $[N/p_i p_j]$ such multiples between 1 and N, we have $\#(A_i \cap A_j) = [N/p_i p_j]^2$. In exactly the same way we see that $\#(A_i \cap A_j \cap A_k) = [N/p_i p_j p_k]^2$, etc. Hence, by the principle of inclusion and exclusion,

$$g(N) = \left[\frac{N}{p_1}\right]^2 + \left[\frac{N}{p_2}\right]^2 + \cdots + \left[\frac{N}{p_m}\right]^2 - \left[\frac{N}{p_1 p_2}\right]^2$$

$$- \left[\frac{N}{p_1 p_3}\right]^2 - \cdots - \left[\frac{N}{p_{m-1} p_m}\right]^2 + \left[\frac{N}{p_1 p_2 p_3}\right]^2 + \cdots$$

$$+ (-1)^{m-1} \left[\frac{N}{p_1 p_2 \cdots p_m}\right]^2.$$

Subtracting this from N^2, we get

$$f(N) = N^2 - \left[\frac{N}{p_1}\right]^2 - \cdots - \left[\frac{N}{p_m}\right]^2 + \left[\frac{N}{p_1 p_2}\right]^2 + \cdots + \left[\frac{N}{p_{m-1} p_m}\right]^2$$

$$- \left[\frac{N}{p_1 p_2 p_3}\right]^2 - \cdots + (-1)^m \left[\frac{N}{p_1 p_2 \cdots p_m}\right]^2.$$

We must now prove that $s_N = [f(N)]/N^2$ tends to a limit as $N \to \infty$. Most of the difficulty in proving this is caused by the brackets in our expression for $f(N)$. Let us denote by $h(N)$ the same expression without the brackets, i.e.,

$$h(N) = N^2 - \frac{N^2}{p_1^2} - \cdots - \frac{N^2}{p_m^2} + \frac{N^2}{p_1^2 p_2^2} + \cdots + (-1)^m \frac{N^2}{p_1^2 p_2^2 \cdots p_m^2}.$$

We can divide both sides by N^2 and then factor the right-hand side, getting

$$\frac{h(N)}{N^2} = \left(1 - \frac{1}{p_1^2}\right)\left(1 - \frac{1}{p_2^2}\right) \cdots \left(1 - \frac{1}{p_m^2}\right) = c_m.$$

As N becomes larger, the number of factors in the product increases, and since these factors are all <1, c_m decreases. Since $c_m > 0$ it tends to a limit[3]

$$s = \lim_{m \to \infty} \left(1 - \frac{1}{p_1{}^2}\right)\left(1 - \frac{1}{p_2{}^2}\right) \cdots \left(1 - \frac{1}{p_m{}^2}\right).$$

We introduce the notation

$$\left(1 - \frac{1}{2^2}\right)\left(1 - \frac{1}{3^2}\right)\left(1 - \frac{1}{5^2}\right)\left(1 - \frac{1}{7^2}\right)\left(1 - \frac{1}{11^2}\right) \cdots$$

or more simply

$$\prod_{p \text{ prim}} \left(1 - \frac{1}{p^2}\right)$$

for this limit, and call it an *infinite product*.

Thus we have shown that $\lim_{N \to \infty} h(N)/N^2 = s$ exists. Our next task is to prove that $\lim_{N \to \infty} [f(N) - h(N)]/N^2 = 0$; we can then add these two equations to get $\lim_{N \to \infty} f(N)/N^2 = s$. We have

$$f(N) - h(N) = \left[\frac{N}{p_1}\right]^2 - \left(\frac{N}{p_1}\right)^2 + \cdots + \left[\frac{N}{p_m}\right]^2 - \left(\frac{N}{p_m}\right)^2 - \left[\frac{N}{p_1 p_2}\right]^2$$
$$+ \left(\frac{N}{p_1 p_2}\right)^2 - \cdots + (-1)^m\left[\frac{N}{p_1 \cdots p_m}\right]^2 - (-1)^m\left(\frac{N}{p_1 \cdots p_m}\right)^2,$$

from which it follows that

$$|f(N) - h(N)| \leqq \left(\frac{N}{p_1}\right)^2 - \left[\frac{N}{p_1}\right]^2 + \cdots + \left(\frac{N}{p_m}\right)^2 - \left[\frac{N}{p_m}\right]^2 + \left(\frac{N}{p_1 p_2}\right)^2$$
$$- \left[\frac{N}{p_1 p_2}\right]^2 + \cdots + \left(\frac{N}{p_1 \cdots p_m}\right)^2 - \left[\frac{N}{p_1 \cdots p_m}\right]^2. \quad (1)$$

The right-hand side of (1) is a sum of nonnegative quantities of the form $(N/r)^2 - [N/r]^2$, where r runs through various products of the primes $p_1, p_2, \ldots, p_m$. The greatest value taken on by r is $p_1 p_2 \cdots p_m$, and this is certainly $\leqq N^m$, since each $p_i \leqq N$. Hence we will increase the right-hand side of (1) if we replace it by the sum of the quantities $(N/r)^2 - [N/r]^2$ where r runs through *all* the integers from 2 to N^m. Therefore

$$|f(N) - h(N)| \leq \left(\frac{N}{2}\right)^2 - \left[\frac{N}{2}\right]^2 + \left(\frac{N}{3}\right)^2 - \left[\frac{N}{3}\right]^2$$
$$+ \left(\frac{N}{4}\right)^2 - \left[\frac{N}{4}\right]^2 + \cdots + \left(\frac{N}{N^m}\right)^2 - \left[\frac{N}{N^m}\right]^2. \quad (2)$$

Now if $r > N$, we have $N/r < 1$, and hence $[N/r] = 0$. If $r \leq N$, we can square both sides of the inequality $[N/r] > N/r - 1$ (since both sides are $\geqq 0$). This gives $[N/r]^2 > (N/r)^2 - 2(N/r) + 1$, from which we obtain

[3] See, e.g., R. Courant, *Differential and Intergal Calculus*, Interscience, New York, 1937, vol. I, pp. 40–41.

$(N/r)^2 - [N/r]^2 < 2N/r - 1 < 2N/r$. Applying these results to the right-hand side of (2), we get

$$|f(N) - h(N)| < \frac{2N}{1} + \frac{2N}{2} + \frac{2N}{3} + \cdots + \frac{2N}{N} + \left(\frac{N}{N+1}\right)^2$$
$$+ \left(\frac{N}{N+2}\right)^2 + \cdots + \left(\frac{N}{N^m}\right)^2.$$

Thus

$$\frac{|f(N) - h(N)|}{N^2} < \frac{2}{N}\left(1 + \frac{1}{2} + \frac{1}{3} + \cdots + \frac{1}{N}\right)$$
$$+ \frac{1}{(N+1)^2} + \frac{1}{(N+2)^2} + \cdots + \frac{1}{(N^m)^2}. \quad (3)$$

Now

$$1 + \frac{1}{2} + \frac{1}{3} + \cdots + \frac{1}{N}$$
$$= \left(1 + \frac{1}{2} + \cdots + \frac{1}{[\sqrt{N}]}\right) + \left(\frac{1}{[\sqrt{N}] + 1} + \cdots + \frac{1}{N}\right).$$

The first parenthesis consists of $[\sqrt{N}]$ terms, all ≤ 1. Therefore its value is $\leq [\sqrt{N}] \leq \sqrt{N}$. The second parenthesis has $N - [\sqrt{N}]$ terms, all $< 1/\sqrt{N}$. Therefore its value is $< (N - [\sqrt{N}])/\sqrt{N} < N/\sqrt{N} = \sqrt{N}$. Combining these estimates, we see that

$$1 + \frac{1}{2} + \frac{1}{3} + \cdots + \frac{1}{N} < 2\sqrt{N}.$$

Next note that

$$\frac{1}{(N+1)^2} + \frac{1}{(N+2)^2} + \frac{1}{(N+3)^2} + \cdots + \frac{1}{(N^m)^2}$$
$$< \frac{1}{N(N+1)} + \frac{1}{(N+1)(N+2)} + \frac{1}{(N+2)(N+3)}$$
$$+ \cdots + \frac{1}{(N^m - 1)N^m}$$
$$= \left(\frac{1}{N} - \frac{1}{N+1}\right) + \left(\frac{1}{N+1} - \frac{1}{N+2}\right) + \left(\frac{1}{N+2} - \frac{1}{N+3}\right)$$
$$+ \cdots + \left(\frac{1}{N^m - 1} - \frac{1}{N^m}\right)$$
$$= \frac{1}{N} - \frac{1}{N^m}$$
$$< \frac{1}{N}.$$

Applying these results to the right-hand side of inequality (3), we find that

$$\frac{|f(N) - h(N)|}{N^2} < \frac{2}{N} \cdot 2\sqrt{N} + \frac{1}{N} = \frac{4}{\sqrt{N}} + \frac{1}{N}.$$

Since the expression on the right tends to 0 as $N \to \infty$, we have

$$\lim_{N \to \infty} \frac{f(N) - h(N)}{N^2} = 0,$$

and the proof is complete.

93. Let

$$t_n = 1 + \frac{1}{2^2} + \frac{1}{3^2} + \cdots + \frac{1}{n^2}.$$

Then $t_n < t_{n+1}$, i.e., the sequence $\{t_1, t_2, t_3, \ldots\}$ is an increasing sequence. Moreover

$$t_n < 1 + \frac{1}{1 \cdot 2} + \frac{1}{2 \cdot 3} + \cdots + \frac{1}{(n-1)n}$$

$$= 1 + \left(1 - \frac{1}{2}\right) + \left(\frac{1}{2} - \frac{1}{3}\right) + \cdots + \left(\frac{1}{n-1} - \frac{1}{n}\right)$$

$$= 2 - \frac{1}{n} < 2.$$

Thus the sequence $\{t_1, t_2, t_3, \ldots\}$ is bounded. Since every bounded increasing sequence has a limit,[4] it follows that $\lim_{n \to \infty} t_n = t$ exists. This means (by the definition of convergence of an infinite series) that

$$1 + \frac{1}{2^2} + \frac{1}{3^2} + \frac{1}{4^2} + \frac{1}{5^2} + \cdots$$

converges to t. Since $\lim_{n \to \infty} (t - t_n) = 0$, we see, furthermore, that

$$\frac{1}{(n+1)^2} + \frac{1}{(n+2)^2} + \frac{1}{(n+3)^2} + \cdots$$

tends to 0 as $n \to \infty$. Now consider the expression

$$\frac{1}{\left(1 - \dfrac{1}{2^2}\right)\left(1 - \dfrac{1}{3^2}\right)\left(1 - \dfrac{1}{5^2}\right) \cdots \left(1 - \dfrac{1}{p_m{}^2}\right)}$$

The geometric series

$$\frac{1}{1 - x} = 1 + x + x^2 + x^3 + \cdots$$

[4] See R. Courant, *op. cit.*, pp. 40–41.

converges for $|x| < 1$; since $0 < 1/p_i{}^2 < 1$, we can therefore write

$$\frac{1}{1 - \dfrac{1}{2^2}} = 1 + \frac{1}{2^2} + \frac{1}{2^4} + \frac{1}{2^6} + \cdots$$

$$\frac{1}{1 - \dfrac{1}{3^2}} = 1 + \frac{1}{3^2} + \frac{1}{3^4} + \frac{1}{3^6} + \cdots$$

$$\vdots$$

$$\frac{1}{1 - \dfrac{1}{p_m{}^2}} = 1 + \frac{1}{p_m{}^2} + \frac{1}{p_m{}^4} + \frac{1}{p_m{}^6} + \cdots .$$

Multiplying these equations together we get

$$\frac{1}{\left(1 - \dfrac{1}{2^2}\right)\left(1 - \dfrac{1}{3^2}\right)\left(1 - \dfrac{1}{5^2}\right) \cdots \left(1 - \dfrac{1}{p_m{}^2}\right)}$$

$$= \left(1 + \frac{1}{2^2} + \frac{1}{2^4} + \frac{1}{2^6} + \cdots\right)\left(1 + \frac{1}{3^2} + \frac{1}{3^4} + \frac{1}{3^6} + \cdots\right)$$

$$\times \left(1 + \frac{1}{5^2} + \frac{1}{5^4} + \frac{1}{5^6} + \cdots\right) \cdots$$

$$\times \left(1 + \frac{1}{p_m{}^2} + \frac{1}{p_m{}^4} + \frac{1}{p_m{}^6} + \cdots\right).$$

We now multiply the infinite series on the right using the distributive law of multiplication.[5] Using the fundamental theorem of arithmetic, which says that every integer >1 can be uniquely expressed as a product of powers of distinct primes,[6] we get

$$\frac{1}{\left(1 - \dfrac{1}{2^2}\right)\left(1 - \dfrac{1}{3^2}\right) \cdots \left(1 - \dfrac{1}{p_m{}^2}\right)} = 1 + \frac{1}{2^2} + \frac{1}{3^2} + \frac{1}{4^2} + \cdots ,$$

where the right-hand side is the sum of the squares of the reciprocals of those integers whose prime decomposition involves only $p_1, \ldots, p_m$.

---

[5] For a proof that this law is valid for infinite series with positive terms, see, e.g., K. Knopp, *Theory and Application of Infinite Series*, London, 1928, pp. 146–147.

[6] For a proof, see, e.g., Hardy and Wright, *op. cit.*, p. 3.

The set of such integers includes the numbers 1, 2, 3, 4, 5, ..., m and also includes some integers $>m$.

Therefore

$$t_m = 1 + \frac{1}{2^2} + \frac{1}{3^2} + \cdots + \frac{1}{m^2} < \frac{1}{\left(1 - \frac{1}{2^2}\right)\left(1 - \frac{1}{3^2}\right) \cdots \left(1 - \frac{1}{p_m^{\,2}}\right)}$$

$$\leq 1 + \frac{1}{2^2} + \frac{1}{3^2} + \cdots = t.$$

Since $\lim_{m \to \infty} t_m = t$, we have

$$\lim_{m \to \infty} \frac{1}{\left(1 - \frac{1}{2^2}\right)\left(1 - \frac{1}{3^2}\right) \cdots \left(1 - \frac{1}{p_m^{\,2}}\right)} = t.$$

But we already proved that

$$\lim_{m \to \infty} \left(1 - \frac{1}{2^2}\right)\left(1 - \frac{1}{3^2}\right) \cdots \left(1 - \frac{1}{p_m^{\,2}}\right) = s.$$

Hence $s = 1/t$.

This fact can be used to compute s to within 0.1. For we have

$$t_n < 1 + \frac{1}{2^2} + \frac{1}{2 \cdot 3} + \frac{1}{3 \cdot 4} + \frac{1}{4 \cdot 5} + \cdots + \frac{1}{(n-1)n}$$

$$= 1 + \frac{1}{4} + \left(\frac{1}{2} - \frac{1}{3}\right) + \left(\frac{1}{3} - \frac{1}{4}\right) + \left(\frac{1}{4} - \frac{1}{5}\right) + \cdots + \left(\frac{1}{n-1} - \frac{1}{n}\right)$$

$$= 1 + \frac{1}{4} + \frac{1}{2} - \frac{1}{n} = \frac{7}{4} - \frac{1}{n} < \frac{7}{4}.$$

Since this holds for all n, $t \leq 7/4$, and therefore $s = 1/t \geq 4/7 = 0.571\ldots$. On the other hand

$$t > 1 + \frac{1}{2^2} + \frac{1}{3 \cdot 4} + \frac{1}{4 \cdot 5} + \cdots + \frac{1}{11 \cdot 12}$$

$$= 1 + \frac{1}{4} + \left(\frac{1}{3} - \frac{1}{4}\right) + \left(\frac{1}{4} - \frac{1}{5}\right) + \cdots + \left(\frac{1}{11} - \frac{1}{12}\right)$$

$$= 1 + \frac{1}{4} + \frac{1}{3} - \frac{1}{12} = \frac{3}{2}.$$

Therefore $s = 1/t < 2/3 = 0.666\cdots$. Combining these estimates we see that $s \approx 0.6$ with an error of less than 0.1.

Remark. It can be shown[7] that $1 + \dfrac{1}{2^2} + \dfrac{1}{3^2} + \dfrac{1}{4^2} + \cdots = \dfrac{6}{\pi^2}$. Thus $s = 6/\pi^2 = 0.608 \cdots$. The identity

$$\left(1 - \frac{1}{2^2}\right)\left(1 - \frac{1}{3^2}\right)\left(1 - \frac{1}{5^2}\right)\left(1 - \frac{1}{7^2}\right) \cdots = \frac{6}{\pi^2}$$

can be used to prove that there are infinitely many primes. For if there were only a finite number of primes, then the left-hand side would be a rational number, whereas π^2 is known to be irrational (see, for example, Hardy and Wright, *An Introduction to the Theory of Numbers*, Oxford, 1960, p. 47). Of course there are much simpler proofs of the infinitude of the primes; the above is an example of cracking open a walnut with a sledgehammer.

94. Denote the first m primes by $p_1, p_2, \ldots, p_m$. We will choose the value of m later. By the remark made at the end of the solution to problem 13, the number of positive integers $\leq N$ which are not divisible by any of the primes $p_1, \ldots, p_m$ is

$$N - \left[\frac{N}{p_1}\right] - \left[\frac{N}{p_2}\right] - \cdots - \left[\frac{N}{p_m}\right] + \left[\frac{N}{p_1 p_2}\right] + \cdots + \left[\frac{N}{p_{m-1} p_m}\right]$$
$$- \left[\frac{N}{p_1 p_2 p_3}\right] - \cdots + (-1)^m \left[\frac{N}{p_1 p_2 \cdots p_m}\right].$$

The primes p such that $p_m < p \leq N$ are not divisible by $p_1, \ldots, p_m$ and are therefore counted in the above expression. Since there are $\pi(N) - m$ such primes p, we have

$$\pi(N) \leq m + N - \left[\frac{N}{p_1}\right] - \cdots - \left[\frac{N}{p_m}\right] + \cdots + (-1)^m \left[\frac{N}{p_1 \cdots p_m}\right]. \quad (1)$$

On the right-hand side of (1) there are $\binom{m}{1} + \binom{m}{2} + \binom{m}{3} + \cdots + \binom{m}{m} = 2^m - 1$ brackets, of which $\binom{m}{1} + \binom{m}{3} + \binom{m}{5} + \cdots = 2^{m-1}$ are preceded by minus signs. Now suppose we remove all the brackets in (1). Since $[a] \leq a$ for any number a, the positive terms will be increased. And since $[a] > a - 1$ each negative term will be decreased by at most 1. Therefore

$$\pi(N) \leq m + 2^{m-1} + N - \frac{N}{p_1} - \cdots - \frac{N}{p_m} + \frac{N}{p_1 p_2}$$
$$+ \cdots + (-1)^m \frac{N}{p_1 \cdots p_m}$$
$$= m + 2^{m-1} + N\left(1 - \frac{1}{p_1}\right)\left(1 - \frac{1}{p_2}\right) \cdots \left(1 - \frac{1}{p_m}\right).$$

[7] See, for example, Knopp, *op. cit.*, p. 237 or Volume 2 of this book.

Since $m \leq 2^{m-1}$, we have

$$\pi(N) \leq 2^m + N\left(1 - \frac{1}{p_1}\right)\left(1 - \frac{1}{p_2}\right) \cdots \left(1 - \frac{1}{p_m}\right),$$

and therefore

$$\frac{\pi(N)}{N} \leq \frac{2^m}{N} + \left(1 - \frac{1}{p_1}\right)\left(1 - \frac{1}{p_2}\right) \cdots \left(1 - \frac{1}{p_m}\right).$$

Now we choose m in such a way that as N tends to infinity, m also tends to infinity, but $2^m/N \to 0$. For example, such a choice of m would be $m = [\log_2 \sqrt{N}]$ (for we then have $2^m/N \leq 2^{\log_2 \sqrt{N}}/\cdot N = \sqrt{N}/N = 1/\sqrt{N}$, which tends to 0 as $N \to \infty$). To complete the proof we need only show that

$$\lim_{m \to \infty} \left(1 - \frac{1}{p_1}\right)\left(1 - \frac{1}{p_2}\right) \cdots \left(1 - \frac{1}{p_m}\right) = 0.$$

By the formula for the sum of a geometric progression, we have

$$\frac{1}{1 - \frac{1}{p}} > \frac{1 - \frac{1}{p^{k+1}}}{1 - \frac{1}{p}} = 1 + \frac{1}{p} + \frac{1}{p^2} + \cdots + \frac{1}{p^k}$$

where k is any positive integer. Therefore

$$\frac{1}{1 - \frac{1}{p_1}} \frac{1}{1 - \frac{1}{p_2}} \cdots \frac{1}{1 - \frac{1}{p_m}} > \left(1 + \frac{1}{p_1} + \frac{1}{p_1^2} + \cdots + \frac{1}{p_1^k}\right)$$

$$\times \left(1 + \frac{1}{p_2} + \frac{1}{p_2^2} + \cdots + \frac{1}{p_2^k}\right) \cdots \left(1 + \frac{1}{p_m} + \frac{1}{p_m^2} + \cdots + \frac{1}{p_m^k}\right).$$

Expanding this last product we obtain a sum of fractions all having numerator 1 but having various integers for their denominators. If k is large enough (e.g., $k = m$ will do), every integer $\leq m$ will appear as one of these denominators, because every integer $\leq m$ can be factored into a product of powers of $p_1, p_2, \ldots, p_m$. Therefore

$$\frac{1}{\left(1 - \frac{1}{p_1}\right)\left(1 - \frac{1}{p_2}\right) \cdots \left(1 - \frac{1}{p_m}\right)} > 1 + \frac{1}{2} + \frac{1}{3} + \frac{1}{4} + \cdots + \frac{1}{m},$$

and so

$$\left(1 - \frac{1}{p_1}\right)\left(1 - \frac{1}{p_2}\right) \cdots \left(1 - \frac{1}{p_m}\right) < \frac{1}{1 + \frac{1}{2} + \frac{1}{3} + \cdots + \frac{1}{m}} \qquad (2)$$

Now as $m \to \infty$, the quantity $1 + \frac{1}{2} + \frac{1}{3} + \cdots + 1/m$ tends to infinity. Because

$$\frac{1}{3} + \frac{1}{4} > \frac{1}{4} + \frac{1}{4} = \frac{1}{2}$$

$$\frac{1}{5} + \frac{1}{6} + \frac{1}{7} + \frac{1}{8} > \frac{1}{8} + \frac{1}{8} + \frac{1}{8} + \frac{1}{8} = \frac{1}{2}$$

$$\frac{1}{9} + \frac{1}{10} + \frac{1}{11} + \frac{1}{12} + \frac{1}{13} + \frac{1}{14} + \frac{1}{15} + \frac{1}{16} > \frac{8}{16} = \frac{1}{2}, \text{ etc.}$$

Thus if $m \geq 2^r$, then

$$1 + \frac{1}{2} + \cdots + \frac{1}{m} > 1 + \overbrace{\frac{1}{2} + \frac{1}{2} + \cdots + \frac{1}{2}}^{r \text{ terms}} = 1 + \frac{r}{2},$$

and this tends to infinity as $r \to \infty$. It follows that $\lim_{m \to \infty} 1/(1 + \frac{1}{2} + \cdots + 1/m) = 0$, and by (2) this implies that

$$\lim_{m \to \infty} \left(1 - \frac{1}{p_1}\right)\left(1 - \frac{1}{p_2}\right) \cdots \left(1 - \frac{1}{p_m}\right) = 0,$$

completing the proof.

Remark. In volume 2 it will be shown that there are two positive constants A and B such that

$$A \frac{N}{\log N} < \pi(N) < B \frac{N}{\log N}.$$

This is a stronger result than what we have just proved, for it implies that

$$\frac{\pi(N)}{N} < \frac{B}{\log N}$$

and clearly $B/\log N \to 0$ as $N \to \infty$.

VIII. EXPERIMENTS WITH A CONTINUUM OF POSSIBLE OUTCOMES

95. The experiment considered in this problem consists in the first person's arriving at some instant x between 12 o'clock and 1 and the second person's arriving at some other instant y in that interval. Thus all the outcomes are defined by pairs of numbers (x,y), where $0 \leq x \leq 1$ and $0 \leq y \leq 1$. Considering these numbers as coordinates of a point in the plane, we can represent the set of all possible outcomes as the totality of all points in or on a square $OABC$ of side 1 (fig. 70). Since x and y are chosen at random between 12 o'clock and 1, the probability that x (or y)

lies in any given interval of the x-axis (resp. y-axis) is equal to the length of that interval (recall that in our case $OA = OC = 1$). Therefore the probability that a point (x,y) lies within any rectangle inside $OABC$ is equal to the area of that rectangle. Since any region can be covered with a network of rectangles, the sum of whose areas differs from the area of the region by an arbitrarily small amount, the probability that the point (x,y) lies in any given region within $OABC$ is equal to the area of that region. (This property could, if desired, be taken as the definition of the concept "at random" which is involved in the statement of this problem.) The favorable outcomes in our problem are those which correspond to points (x,y) for which $|x - y| \leq 1/4$. These points make

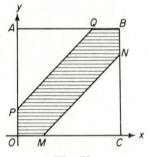

Fig. 70

up the shaded region in fig. 70 bounded by the straight lines $MN(x - y = 1/4)$ and $PQ(y - x = 1/4)$. Here $OM = BN = OP = BQ = 1/4$; consequently, area $MCN =$ area $PAQ = 1/2 \cdot (3/4)^2 = 9/32$, and the required probability is area $PQBNMO = 1 -$ area $\cdot MCN -$ area $PAQ = 1 - 18/32 = 7/16$.

96. *First solution.* Let AB be our rod and K and L the two break points (fig. 71a). Denote the length of AB by l. All possible outcomes of the experiment considered in this problem are determined by the locations of the points K and L, in other words, by the two numbers $AK = x$ and $AL = y$, each of which is to be chosen at random between 0 and l. Considering these numbers as the coordinates of a point in the plane, we can represent the set of all possible outcomes as the totality of all points in or on a square $OMNP$ of side l (fig. 71b). As in problem 95, the probability that the point (x,y) lies in any given region S of this square is equal to the ratio of the area of S to the area of the entire square. Therefore, the problem is to determine the area of the part of the square $OMNP$ which consists of the points corresponding to favorable outcomes to the experiment.

In order that the three pieces into which the rod is broken can be arranged to form a triangle, it is necessary and sufficient that the length of each of these three pieces be less than the sum of the lengths of the other two pieces. Since the sum of the lengths of the pieces is l, this condition is equivalent to the requirement that each of the three pieces be less than $l/2$ in length. Now consider under what circumstances this condition could fail to be satisfied.

1. The length of the leftmost of the three pieces will exceed $\frac{1}{2}l$ if x and y are both greater than $\frac{1}{2}l$. Such outcomes correspond to points located within the small square $YRZN$ of width $\frac{1}{2}l$ which occupies the upper right-hand corner of the square $OMNP$ (fig. 71b).

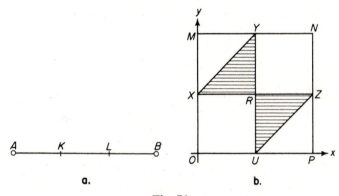

Fig. 71

2. The length of the rightmost of the three pieces will exceed $\frac{1}{2}l$ if x and y are both less than $\frac{1}{2}l$. Such outcomes correspond to the points located within the small square $XRUO$ of width $\frac{1}{2}l$ which occupies the lower left-hand corner of the square $OMNP$ (fig. 71b).

3. The length of the middle segment will be greater than $\frac{1}{2}l$ if x and y satisfy the inequality $y - x > \frac{1}{2}l$ or the inequality $x - y > \frac{1}{2}l$. The set of all points (x,y) for which $y - x = \frac{1}{2}l$ is represented in the figure by the line XY and the set of all points for which $x - y = \frac{1}{2}l$ by the straight line ZU. The points for which $y - x > \frac{1}{2}l$ fill the triangle XMY and the points for which $x - y > \frac{1}{2}l$ fill the triangle UPZ. Thus the unfavorable outcomes of the third type correspond to the points within the triangles XMY and UPZ.

Finally, the favorable outcomes correspond to the points within the shaded part of the square $OMNP$ (fig. 71b). Since the area of this part is one-fourth of the area of the large square, the desired probability is 1/4.

Second solution. In the first solution we characterized the outcomes of the experiment in terms of the numbers $x = AK$ and $y = AL$. Instead of these we could have used the numbers $x = AK$ and $z = KL$, where K is the leftmost of the two break points (see fig. 71a). It is clear that $x + z \leqq l$ (since l is the length of the entire rod and $x + z$ is the sum of the lengths of two of the three pieces); therefore, the set of all possible outcomes will be represented here by the points in or on the triangle OST bounded by the coordinate axes and the straight line $x + z = l$.

It is easy to see that the probability in the case of such a choice of coordinates is again proportional to the area. The set of all favorable outcomes is defined by the inequalities $x < \frac{1}{2}l$, $z < \frac{1}{2}l$, $x + z > \frac{1}{2}l$. By drawing the straight lines $x = \frac{1}{2}l$, $z = \frac{1}{2}l$, $x + z = \frac{1}{2}l$ (these are

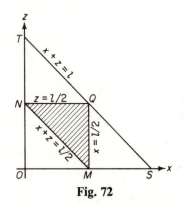

Fig. 72

the lines QM, QN, and MN in fig. 72), we see that the favorable outcomes correspond to the points inside the shaded triangle in fig. 72. Since the area of this triangle is one-fourth of the area of the entire triangle OST, we again obtain the value 1/4 for the probability in question.

Remark. It is instructive to compare this result with problems 27 and 30. In problem 27 we showed that the number of solutions of the equation

$$x + y + z = n \tag{1}$$

in positive integers is $(n - 1)(n - 2)/2$. In the first part of the solution to problem 30 we showed that the number of these solutions satisfying the inequalities

$$x + y < z, \qquad x + z < y, \qquad y + z < x \tag{2}$$

is $(n - 2)(n - 4)/8$ for even n, and $(n - 1)(n + 1)/8$ for odd n. Hence the probability p_n that a solution of (1) chosen at random satisfies (2) is $(n - 4)/4(n - 1)$ for even n, and $(n + 1)/4(n - 2)$ for odd n. In either case p_n approaches 1/4 as $n \to \infty$. Now p_n can be interpreted geometrically as follows. Let the rod AB of length l be divided into n equal pieces by points $A_0 = A$,

$A_1, A_2, \ldots, A_n = B$. Then to every solution of (1) there corresponds a division of the rod AB into three non-empty pieces, the break point being chosen from $A_1, A_2, \ldots, A_{n-1}$. The solutions satisfying (2) correspond to those divisions with the property that a triangle can be formed from the three pieces. Therefore if the rod AB is broken at random into three non-empty pieces, where the break points are among $A_1, A_2, \ldots, A_{n-1}$, then p_n is the probability that a triangle can be formed from the pieces. (Here the term "at random" means that all solutions of (1) in nonnegative integers are to be considered as equally likely. This is not quite the same as saying that the break points are chosen independently and uniformly from $A_1, \ldots, A_{n-1}$.) The fact that $p_n \to 1/4$ as $n \to \infty$ means that in a certain sense, problem 95 can be thought of as a limiting case of the "discrete" problem, when the division points are restricted as above. At the same time, the solution to the present problem is appreciably simpler than that of problems 27 and 30: instead of making complicated combinatorial computations, here we need only determine the ratio of the areas of two similar triangles. (A similar situation is encountered in many considerably deeper questions: in modern probability theory the passage from finite cases to the continuous case plays an important role and often allows the theory to be simplified considerably, by the elimination of complicated combinatorial computations and estimates.)

97. This problem is a generalization of the preceding one. For problem 96 can be reformulated as follows: what is the probability that none of three pieces into which a rod is broken at random will exceed half the total length of the rod? In this problem the length $\frac{1}{2}l$ is replaced by an arbitrary length a.

As in the first solution of problem 96, we will characterize all possible outcomes of the experiment in question in terms of two numbers $AK = x$ and $AL = y$, where K and L are the break points and A is the left end of the rod. Then the set of all possible outcomes can be represented as the set of all points in or on a square $OMNP$ of side l. The probability that a point (x,y) lies within any region R of this square is equal to the ratio of the area of R to the area of the entire square.

Let us determine what part of the square is filled by the points corresponding to unfavorable outcomes of the experiment. First of all, it is obvious that for $a < l/3$ all outcomes of the experiment are unfavorable: at least one of the pieces will have to have length greater than or equal to $l/3$. Let us now consider separately the cases $l/3 \leq a \leq l/2$ (fig. 73a) and $a \geq l/2$ (fig. 73b).

In order that the leftmost piece of the rod have length greater than a, it is necessary and sufficient that $x > a$, $y > a$; the points which satisfy these conditions make up the square $YRZN$ of width $l - a$ in figs. 73a and 73b. Similarly, in order that the rightmost piece have length greater than a, it is necessary and sufficient that $x < 1 - a$, $y < 1 - a$; these points (x,y) make up the square $XQUO$ of width $l - a$. Finally, the middle piece of the rod will have length greater than a if and only if either $x - y > a$ or $y - x > a$; the first condition is satisfied by the points of the triangle *VMW* and the second by the points of the triangle *SPT*.

Now we need merely compute the area of the unshaded portion of the squares in figs. 73a and 73b. In the first case (fig. 73a) this will consist of two isosceles triangles whose legs have length $PS - US - YT = l - a - 2(l - 2a) = 3a - l$; therefore, the area of the unshaded portion is $2 \cdot \frac{1}{2}(3a - l)^2 = (3a - l)^2$. In the second case (fig. 73b) we must subtract from the area of the square the area of two squares of width $l - a$ and two isoceles right triangles whose legs have length $l - a$;

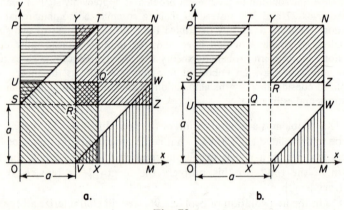

a.   b.

Fig. 73

therefore, the area of the unshaded portion is $l^2 - 3(l - a)^2$. Dividing these areas by l^2 (the area of the entire square), we find that the probability that none of the three pieces has length greater than a is

$$0 \quad \text{if} \quad 0 \leqq a < l/3,$$

$$\left(3\frac{a}{l} - 1\right)^2 \quad \text{if} \quad l/3 \leqq a \leqq l/2,$$

$$1 - 3\left(1 - \frac{a}{l}\right)^2 \quad \text{if} \quad \tfrac{1}{2}l \leqq a \leqq l.$$

98. Since after a random choice of the three points, the circle can always be rotated into a position in which one of the points (say, *A*) is at a preassigned position, we can assume that *A* is fixed to begin with and that only two points are chosen at random. The locations of the points *B* and *C* will then be determined by the lengths of the arcs *AB* and *AC* reckoned in a specific direction (say, counterclockwise from *A*). Now cut the circle at the point A and straighten it out into a line segment AA' (fig. 74; the fact that both ends of this segment represent the same point of the circle is immaterial). For the triangle ABC to be obtuse (that is, for the outcome to be unfavorable), it is necessary and sufficient that one of the three arcs into which the vertices divide the circle is more than a

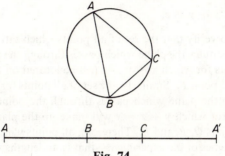

Fig. 74

semicircle. Hence after passing from the circle to the segment AA' our problem assumes the following form: two points B and C are selected at random on the segment AA': what is the probability that none of the three pieces into which these points divide the segment is longer than half the segment? But we showed in problem 96 that this probability is 1/4.

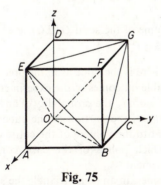

Fig. 75

99. Here all possible outcomes to the experiment are determined by a triple of numbers (x,y,z), each number being chosen at random between 0 and l, where l is the length of the rod under consideration. By considering these numbers as coordinates of a point in space, we can represent the possible outcomes by the points of a cube of side l (fig. 75). Then the probability that the point (x,y,z) lies within any given region of the cube is equal to the ratio of the volume of that region to the volume of the entire cube.

Let us now determine the location of the points which correspond to favorable outcomes of the experiment. In order to form a triangle from three segments of lengths x, y, and z, it is necessary and sufficient that they satisfy the inequalities

$$x + y > z, \qquad x + z > y, \qquad y + z > x. \tag{1}$$

But it is easy to verify that the set of all points which satisfy the equation $x + y = z$ constitutes the plane which passes through the points O, E, and G, and all points for which $x + y > z$ will be located on the same side of this plane as the point F. Similarly, the set of all points for which $x + z = y$ will make up the plane which passes through the points O, B, and G; and all points for which $y + z = x$ will make up the plane which passes through the points O, B, and E. Therefore, all points which correspond to favorable outcomes of the experiment—that is, to lengths which satisfy the inequalities (1)—will be located within the region $OBGEF$ bounded by the three planes indicated above and three faces of the cube. Note now that the volume of the cube differs from the volume of the solid $OBGEF$ by three times the volume of the triangular pyramid $EOBA$ (since the triangular pyramids $EOBA$, $GOED$, and $GOBC$ are congruent). The volume of $EOBA$ is

$$\tfrac{1}{3}EA \cdot \text{area } OAB = \tfrac{1}{3}l \cdot \tfrac{1}{2}l^2 = \frac{l^3}{6} \,;$$

consequently, the volume of the solid $OBEGF$ is $l^3 - 3\dfrac{l^3}{6} = \dfrac{l^3}{2}$; and the required probability is $\tfrac{1}{2}\,l^3/l^3 = \tfrac{1}{2}$.

100. The experiment here is the same as in the preceding problem. Again we represent the possible outcomes of the experiment by points in or on a cube $OABCDEFG$ of width l; the probability of any event will be equal to the ratio of the volume of the portion of the cube corresponding to the outcomes of that event to the volume of the entire cube. Hence we have only to determine which points of the cube correspond to favorable outcomes of the experiment.

By the law of cosines, in an acute triangle the square of the length of any side is less than the sum of the squares of the lengths of the other two sides; on the other hand, in an obtuse triangle the square of the length of the side opposite the obtuse angle is greater than the sum of the squares of the other two sides, and in a right triangle the square of the hypotenuse equals the sum of the squares of the legs. Consequently, in order that it be possible to form an acute triangle from three segments, it is necessary and sufficient that the square of the length of each of the segments be less than the sum of the squares of the lengths of the other two segments.[1] Therefore, the difference between this problem and the preceding one is only that now the favorable outcomes are defined not by the inequalities (1) of problem 99, but by the inequalities

$$x^2 + y^2 > z^2, \qquad x^2 + z^2 > y^2, \qquad y^2 + z^2 > x^2. \qquad (2)$$

The equation $x^2 + y^2 = z^2$ means that the distance $MQ = \sqrt{x^2 + y^2}$ from the point $M = (x,y,z)$ to the z-axis equals the segment OQ of the z-axis (fig. 76a). It is clear from this that all points for which $x^2 + y^2 = z^2$ will lie on the surface of a cone having the line OD as axis and an angle of $45°$ between the axis and the generator; the points for which $z^2 > x^2 + y^2$

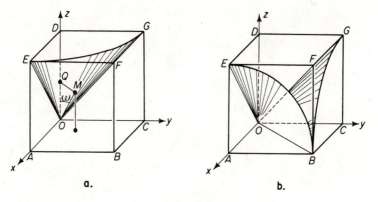

Fig. 76

(these points correspond to unfavorable outcomes of the experiment) will lie within this cone. Similarly, the inequalities $y^2 > x^2 + z^2$ and $x^2 > y^2 + z^2$ define the interiors of two cones having the lines OC and OA as axes (fig. 76b). A quarter of each of these cones is located within the cube $OABCDEFG$. They do not cross; the altitude of each of them is l, and the radius of the base of each is also l (since the angle ω between the axis and the generator equals $45°$). The volume of a quarter of such a cone is

$$\frac{1}{4} \cdot \frac{1}{3} \pi l^2 \cdot l = \frac{\pi l^3}{12} ;$$

consequently, the volume of the part of the cube which corresponds to the unfavorable outcomes of the experiment is $3(\pi l^3/12) = \pi l^3/4$, and the

---

[1] This condition guarantees that none of the segments will exceed the sum of the other two (that is, that one will be able to form a triangle from these segments). For if, for example, $x^2 < y^2 + z^2$, then certainly $x^2 < y^2 + 2yz + z^2 = (y + z)^2$, which means that $x < y + z$.

volume of the part of the cube which corresponds to the favorable outcomes is $l^3 - \pi l^3/4$.

It follows from this that the desired probability is

$$\frac{l^3 - \dfrac{\pi l^3}{4}}{l^3} = 1 - \frac{\pi}{4} = 0.2146\ldots.$$

ANSWERS AND HINTS

1. 3.

2. 7.

3. 7.

4. $5 + \binom{5}{2} = 15$.

5. 8 (in certain special cases there can be less than 8).

6. 30 ways.

7. 22,754,499,243,840 ways $\left(\text{this number equals } \dfrac{33!}{(11!)^3 3!}\right)$.

8. $\binom{16}{6} = 8008$ ways.

9. The number of locks required is $\binom{11}{5} = 462$; the number of keys held by each committee member is $\binom{10}{5} = 252$.

10. 800.

11a. There are more numbers in which a 1 occurs.

  b. 175,308,642.

12. Show that if an element e is contained in exactly $h \geq 1$ of the sets, then e is counted $\binom{h}{1} - \binom{h}{2} + \binom{h}{3} - \cdots + (-1)^h \binom{h}{h}$ times on the right-hand side of the formula. Apply the binomial theorem to show that this expression is equal to 1.

13a. 686.

  b. 457.

14. 288.

15. 7142. Prove that the remainders from dividing $2^x - x^2$ by 7 repeat periodically in groups of 21 as x varies from 1 to 10,000.

16. 10,153. Prove that $x^2 + y^2$ is divisible by 7 only when x and y are both divisible by 7.

17. 139 ways. Compute the number of factorizations, counting factorizations which differ in the order of the factors as different (there will be 784 such factorizations) and then take into account the fact that some of these 784 factorizations must be considered as the same.

18. The number of divisors is 60. The sum of the divisors is 62,868.

19. 473.

20. The coefficient of x^{18} is zero; the coefficient of x^{17} is 3420.

21. 12 ways.

22. $[n/5] + 1$ ways.

23a. $N\left(\dfrac{(n+3)^2}{12}\right)$ ways (regarding the notation, see p. 6). To prove this formula, use the result of problem 22 and the fact that

$$\left[\frac{m}{2}\right] = \frac{m}{2} - \frac{1}{4} + \frac{(-1)^m}{4}$$

(for m an integer).

  b. $N\left(\dfrac{(n+4)^2}{20}\right)$ ways.

24. 4562 ways. In solving this problem it is convenient to make use of the result of problem 23b.

25. $[n/2]$ ways.

26. 19,801 solutions.

27. $\dfrac{(n-1)(n-2)}{2}$ ways.

28a. $N\left(\dfrac{(n+3)^2}{12}\right)$ ways.

  b. $N\left(\dfrac{n^2}{12}\right)$.

29. $\dfrac{n^2-1}{8}$ solutions for odd n and $\dfrac{(n+8)(n-2)}{8}$ solutions for even n.

30. $N\left(\dfrac{n^2+6n}{48}\right)$ triangles for odd n, and $N\left(\dfrac{n^2}{48}\right)$ triangles for even n.

In solving this problem one should use the result of problem 29 and consider separately the 12 cases corresponding to the 12 possible remainders of n when divided by 12.

31a. $\dbinom{n-1}{m-1}$.

  b. $\dbinom{n+m-1}{m-1}$.

32a. Let n be the sum of k_1 terms equal to 1, k_2 terms equal to 2, ... and k_p terms equal to p, where $k_1 + k_2 + \cdots + k_p \leq m$ (some of the k_i's may be 0). Put $y_1 = k_1 + k_2 + \cdots + k_p, y_2 = k_2 + k_3 + \cdots + k_p, \ldots,$ $y_p = k_p$. Then $n = y_1 + y_2 + \cdots + y_p$ and all the y_j's are $\leq m$. The theorem can be derived easily from this.

  b. Let

$$n = x_1 + x_2 + \cdots + x_m$$

be a representation of the number n as a sum of m distinct terms arranged in increasing order; then

$$n - \frac{m(m+1)}{2} = (x_1 - 1) + (x_2 - 2) + \cdots + (x_m - m)$$

$$= k_0 \cdot 0 + k_1 \cdot 1 + k_2 \cdot 2 + \cdots + k_p \cdot p$$

is a partition of number $n - \dfrac{m(m+1)}{2}$. Denoting $k_1 + k_2 + k_3 + \cdots + k_p$ by y_1, $k_2 + k_3 + \cdots + k_p$ by y_2, ..., and k_p by y_p, we obtain

$$y_1 + y_2 + \cdots + y_p = n - \frac{m(m+1)}{2}.$$

33a. Represent each term as the product of a power of 2 and an odd number, and collect the terms which have the same odd factor.

b. Let the term 1 occur s_1 times, and term 2 s_2 times, etc., in a representation of the number n as a sum of terms not divisible by k. Write the numbers $s_1, s_2, \ldots$ in "k-ary form."

34a. The greatest number of rooks is n; the number of arrangements is $n!$.

b. The smallest number of rooks is n; the number of arrangements is $2n^n - n!$.

35a. 14; $2n - 2$.

b. 8; n.

36. Use the fact that for even n the union of the white squares is congruent to the union of the black squares.

37a. Use the result of problem 35a and compute the number of squares which each of the eight bishops controls.

b. The number of arrangements is 2^n. To prove this, use the result of part a.

38a. $108^2 = 11,664$.

b. $672^2 = 451,584$.

c. $24 \cdot 2964 = 71,136$.

d. $\left[(2k)! \dfrac{4k+1}{2} \right]^2$ if $n = 4k$

$(2k)!^2 \dfrac{16k^3 + 24k^2 + 11k + 1}{2}$ if $n = 4k + 1$.

$(2k)!^2 (4k^2 + 5k + 2)^2$ if $n = 4k + 2$.

$(2k)! (2k+1)! (16k^4 + 56k^3 + 67k^2 + 33k + 6)$ if $n = 4k + 3$.

39a. 16.

b. $\left[\dfrac{n+1}{2} \right]^2$.

40a. 9.

b. $\left[\dfrac{n+2}{3} \right]^2$.

41a. 8 queens.

    b. One queen for $n = 1$ and $n = 2$, two queens for $n = 3$, n queens for $n \geq 4$. In order to prove that n queens can be arranged in the required fashion on an $n \times n$ chessboard (for arbitrary $n \geq 4$), it is sufficient to consider only the case of even n and to produce for this case an admissible arrangement of the n queens in which the centermost positive diagonal is left empty. (The case of odd n can then be handled by adding an extra row and column and putting the extra queen on the corner square.) As regards even n, it is convenient to consider first the case of n giving a remainder of 0 or 4 upon division by 6 and then the case of n giving a remainder of 2 upon division by 6.

42a. 32 knights.

    b. Two arrangements.

43a. Into $(n + 1)^2$ parts.

    b. Into $3n^2 + 3n + 1$ parts.

44a. Into $\dfrac{n^2 + n + 2}{2}$ parts.

    b. Into $n^2 - n + 2$ parts.

    In solving the problem it is convenient to consider in turn by how much the number of parts is increased after drawing the second, third, ..., and n-th line (or circle).

45a. Into $\dfrac{n^3 + 5n + 6}{6}$ parts. Use the result of problem 44a.

    b. Into $\dfrac{n(n^2 - 3n + 8)}{3}$ parts. Use the result of problem 44b.

46. In $\dfrac{n(n - 1)(n - 2)(n - 3)}{23} = \binom{n}{4}$ points.

47. Into $\dfrac{(n - 1)(n - 2)(n^2 - 3n + 12)}{24}$ parts.

48a. $1296 = 36^2$.

    b. $\dfrac{n^2(n + 1)^2}{4}$.

49a. 204.

    b. $1^2 + 2^2 + 3^2 + \cdots + n^2 = \dfrac{n(n + 1)(2n + 1)}{6}$.

50. $\dfrac{n(n - 1)(n - 2)(n^3 + 18n^2 - 43n + 60)}{720}$ triangles. This expression is obtained as the sum $\binom{n}{3} + 4\binom{n}{4} + 5\binom{n}{5} + \binom{n}{6}$.

51. The required number of k-gons is

$$\frac{n(n-k-1)!}{k!\,(n-2k)!} = \frac{n}{k}\binom{n-k-1}{k-1}.$$

52a. The number of triangles is $n-2$.

  b. The number of diagonals is $n-3$.

53a. In 132 ways. In solving this problem it is convenient to determine successively the number of decompositions into triangles of a convex quadrilateral, pentagon, hexagon, heptagon, and octagon.

  b. The required number T_n of ways is $\dfrac{1 \cdot 3 \cdot 5 \cdots (2n-5)}{(n-1)!}\, 2^{n-2}$

To derive this formula we need only to prove that

$$T_{n+1} = \frac{2(2n-3)}{n}\, T_n.$$

One of the ways of obtaining this relation is the following: We express T_n in terms of T_{n-1}, $T_{n-2}, \ldots$, T_3 in two different ways and derive the required relation by equating the two formulas obtained.

54. The required number F_n of ways is $\dfrac{1 \cdot 3 \cdot 5 \cdots (2n-1)}{(n+1)!}\, 2^n$.

To derive this formula it suffices to show that $F_n = T_{n+2}$ (see the hint to problem 53b).

55a. In $\dfrac{n^p - n}{p} + n$ ways.

  b. The answer to part a must be an integer.

56a. $\dfrac{1}{2}\left\{\dfrac{(p-1)! + 1}{p} + p - 4\right\}$ self-intersecting p-gons.

  b. The answer to part a must be an integer.

57a. 2^n.

  b. 0 if $n > 1$, 1 if $n = 1$.

  c. $\dfrac{2^{n+1} - 1}{n+1}$.

  d. $2^{n-1}n$.

  e. 0 if $n > 1$, 1 if $n = 1$.

  f. $(-1)^m\binom{n-1}{m}$ for $m < n$ and 0 for $m = n$.

  g. $\binom{n+m+1}{k+1} - \binom{n}{k+1}$ for $k < n$ and $\binom{n+m+1}{n+1}$ for $k = n$.

  h. 1 for $n = 3k$, 0 for $n = 3k + 1$, and -1 for $n = 3k + 2$.

  i. 2^{2n}.

  j. $\binom{2n}{n}$.

k. $(-1)^{n/2}\binom{n}{n/2}$ for even n, and 0 for odd n.

l. $\binom{n+m}{k}$.

58a. 2^{n-1} if $n > 0$, 1 if $n = 0$.

b. 2^{n-1}.

The solutions to these two problems follow easily from the results of problems 57a and b.

c. $2^{n-2} + 2^{(n-2)/2}$ for $n = 8k$, $k > 0$; $2^{n-2} + 2^{(n-3)/2}$ for $n = 8k + 1$; 2^{n-2} for $n = 8k \pm 2$; $2^{n-2} - 2^{(n-3)/2}$ for $n = 8k \pm 3$; and $2^{n-2} - 2^{(n-2)/2}$ for $n = 8k + 4$.

To solve this and the following three problems make use of the expansions of $(1 + i)^n$ and $(1 - i)^n$ according to the binomial theorem.

d. 2^{n-2} for $n = 8k$ or $n = 8k + 4$, $2^{n-2} + 2^{(n-3)/2}$ for $n = 8k + 1$ or $n = 8k + 3$, $2^{n-2} + 2^{(n-2)/2}$ for $n = 8k + 2$, $2^{n-2} - 2^{(n-3)/2}$ for $n = 8k - 1$ or $n = 8k - 3$, and $2^{n-2} - 2^{(n-2)/2}$ for $n = 8k - 2$.

e. $2^{n-2} - 2^{(n-2)/2}$ for $n = 8k$, $2^{n-2} - 2^{(n-3)/2}$ for $n = 8k \pm 1$, 2^{n-2} for $n = 8k \pm 2$, $2^{n-2} + 2^{(n-3)/2}$ for $n = 8k \pm 3$, and $2^{n-2} + 2^{(n-2)/2}$ for $n = 8k + 4$.

f. 2^{n-2} for $n = 8k$ or $n = 8k + 4$, $2^{n-2} - 2^{(n-3)/2}$ for $n = 8k + 1$ or $n = 8k + 3$, $2^{n-2} - 2^{(n-2)/2}$ for $n = 8k + 2$, $2^{n-2} + 2^{(n-3)/2}$ for $n = 8k - 1$ or $n = 8k - 3$, and $2^{n-2} + 2^{(n-2)/2}$ for $n = 8k - 2$.

g. $(2^n + 2)/3$ for $n = 6k$, $(2^n + 1)/3$ for $n = 6k \pm 1$, $(2^n - 1)/3$ for $n = 6k \pm 2$, and $(2^n - 2)/3$ for $n = 6k + 3$.

To solve this and the following two problems, make use of the expansions of $(1 + \omega)^n$ and $(1 + \omega^2)^n$ according to the binomial theorem, where $\omega = (-1 + i\sqrt{3})/2$ and $\omega^2 = (-1 - i\sqrt{3})/2$.

h. $(2^n - 1)/3$ for $n = 6k$ or $n = 6k - 2$, $(2^n + 1)/3$ for $n = 6k + 1$ or $n = 6k + 3$, $(2^n + 2)/3$ for $n = 6k + 2$, and $(2^n - 2)/3$ for $n = 6k - 1$.

i. $(2^n - 1)/3$ for $n = 6k$ or $n = 6k + 2$, $(2^n - 2)/3$ for $n = 6k + 1$, $(2^n + 1)/3$ for $n = 6k + 3$ or $n = 6k - 1$, and $(2^n + 2)/3$ for $n = 6k - 2$.

59. Use mathematical induction.

60a. $\binom{m+n}{k}$ (see problem 57*l*).

b. $(n - m - 1)(n - m - 2) \cdots (n - m - k)/k!$. For $n - m - 1 \geq k$, this expression equals $\binom{n - m - 1}{k}$.

61c. $\binom{n+m}{k}$

62a. 1; 1000; 499,500

b. $\binom{1000}{k} = \dfrac{1000!}{k!\,(1000 - k)!}$ people arrive at the k-th intersection of the 1000th row.

63. Make use of the fact that $B_n{}^k$ is the coefficient of x^k in the expansion of the expression $(1 + x + x^2)^n$.

64. $0.6561 = (0.9)^4$. In solving this problem it is convenient to treat all the license numbers as four-digit numbers by adding initial zeros where necessary and by replacing the number 10,000 by 0000.

65a. $1/360 \approx 0.003$.

   b. $12/360 = 1/30 \approx 0.033$.

66. $76/30{,}240 = 19/7560 \approx 0.0025$. In the solution take into account the fact that $495 = 5 \cdot 9 \cdot 11$, and use criteria for divisibility by 5, by 9, and by 11.

67. 0.2.

68a. $\dfrac{12!}{12^{12}} \approx 0.000054$.

   b. $\dfrac{66(2^6 - 2)}{12^6} \approx 0.00137$.

69a. $\dfrac{9 \cdot 8 \cdot 7 \cdot 2^6}{1 \cdot 2 \cdot 3 \cdot 3^9} = \dfrac{1792}{6561} \approx 0.273$.

   b. $\dfrac{9!}{(3!)^3 \cdot 3^9} = \dfrac{560}{6561} \approx 0.085$.

   c. $\dfrac{9!}{2 \cdot 4! \cdot 3^9} = \dfrac{280}{729} \approx 0.384$.

70a. $\frac{4}{9}$.

   b. $\frac{1}{6}$.

   c. $\frac{10}{21}$.

71. Winning three games out of four is more probable than winning five games out of eight (the first probability is 1/4 and the second is 7/32).

72a. $\dfrac{\dbinom{n}{r}\dbinom{m}{k - r}}{\dbinom{n + m}{k}}$

   b. $\dbinom{n + m}{k}$. Use the fact that the sum of the probabilities of no white balls being drawn, of exactly one being drawn, of exactly two being drawn, . . . , and of exactly k being drawn equals 1.

73a. $\dfrac{\dbinom{2n - k}{n}}{2^{2n-k}}$

   b. 2^{2n}. This result can be derived from part a in the same way that problem 72b follows from 72a.

74. $1/2$. The simplest way to solve this problem is to show directly that the number of favorable outcomes is equal to the number of unfavorable outcomes (without computing this number).

75a. 5/9. (One can treat the situation considered in this problem as having nine equally likely possible outcomes and show that five of them are favorable.)

b. 19/27.

c. $1 - (\frac{2}{3})^n$. Here it is easier to compute the number of cases in which all n hunters miss.

76. $\dfrac{95}{144} \approx 0.66$.

77. 13/41. In solving this problem it is essential to note that of the 81 equally likely possible outcomes if A, B, C, and D made independent statements, only 41 are actually possible in this case, by virtue of the special character of the statements.

78a. $8/15 \approx 0.53$.

b. $\dfrac{2 \cdot 4 \cdot 6 \cdots (2n - 2)}{1 \cdot 3 \cdot 5 \cdots (2n - 1)}$.

79a. $\dfrac{2^n (n!)^2}{(2n)!}$.

b.
$$\begin{cases} 0 & \text{for odd } n, \\[2mm] \dfrac{(2k)!^3}{(4k)!\,(k!)^2} & \text{for even } n = 2k. \end{cases}$$

80a. $1 - \dfrac{1}{2!} + \dfrac{1}{3!} - \dfrac{1}{4!} + \cdots + (-1)^{n+1} \dfrac{1}{n!}$.

b. As $n \to \infty$, this probability approaches $1 - \dfrac{1}{e} \approx 0.632$ ($e \approx 2.718$) is the base of the system of natural logarithms.

81a. $\dbinom{m}{0} 1^p - \dbinom{m}{1}\left(1 - \dfrac{1}{m}\right)^p + \dbinom{m}{2}\left(1 - \dfrac{2}{m}\right)^p - \cdots + (-1)^{m-1}$
$$\cdot \dbinom{m}{m-1}\left(1 - \dfrac{m-1}{m}\right)^p.$$

b. $\dfrac{\dbinom{m}{r}}{m^p}\left\{\dbinom{r}{0} r^p - \dbinom{r}{1}(r-1)^p + \cdots + (-1)^{r-1}\dbinom{r}{r-1} 1^p\right\}$.

c. $\dbinom{m}{1} 1^p - \dbinom{m}{2} 2^p + \dbinom{m}{3} 3^p - \cdots + (-1)^{m-1}\dbinom{m}{m} m^p$
$$= \begin{cases} 0 & \text{if } p < m \\[2mm] (-1)^{m-1} m! & \text{if } p = m \end{cases}.$$

82. $\dfrac{439{,}792}{3{,}628{,}800} \approx 0.12$. The total number of possible outcomes to the experiment is $9! \cdot 10!$ The number of unfavorable outcomes is $\dbinom{10}{1}a_1 - \dbinom{10}{2}a_2 + \dbinom{10}{3}a_3 - \cdots + \dbinom{10}{9}a_9 - \dbinom{10}{10}a_{10}$, where

$$a_k = \begin{cases} (10-k)!\,(10-k-1)!\,(20-2k)(20-2k+1)\cdots(20-k-1) & \\ \hspace{4cm} \text{for } 1 \le k \le 9, \\[2mm] 2 \cdot 9!. & \text{for } k = 10. \end{cases}$$

83a. $(n - m + 1)/(n + 1)$ for $m \le n$; 0 for $m > n$. This answer can be obtained in various ways.

First solution. There are a total of $\dbinom{n+m}{m}$ equally likely possible outcomes to our experiment. It is convenient to represent these outcomes as the $\dbinom{n+m}{m}$ shortest paths joining the intersections $(0,0)$ and (n,m) of a network of roads (see p. 18). Using this representation one can show for $m \le n$ the number of unfavorable outcomes equals the number of paths joining the intersections $(-1,1)$ and (n,m), that is $\dbinom{n+m}{m-1}$; the answer to the problem follows easily from this.

Second solution. Given the answer to the problem, it is not hard to verify its validity by mathematical induction.

Third solution. Consider the $n + m$ arrangements which are obtained from a given arrangement by successively moving the first customer to the end of the line. Show that for $n > m$ exactly $n - m$ of these $n + m$ arrangements have the property that in front of each customer there are more people with fives than people with only tens. It will not be hard to obtain the answer to the problem from this.

b. $1 - \dfrac{m(m-1)\cdots(m-p)}{(n+1)(n+2)\cdots(n+p+1)}$ for $n + p \ge m \ge p + 1$; 0 for $m > n + p$; 1 for $m \le p$. The solution to this problem can be carried out analogously to the first or second solution of part a.

c. $(n - 2m + 1)/(n + 1)$ for $n \ge 2m$ and 0 for $n < 2m$.

The solution of this problem can be carried out similarly to the first, second, or third solution of part a, the simplest way being the solution analogous to the third solution of part a.

84a. In a division of the $2n$ points on the circle into n pairs, n points will be the first points of the pairs and the other n points the second points. Show that in order for the n chords determined by these pairs not to intersect it is necessary and sufficient that (numbering the points in the order in which they occur around the circle) each point be preceded by at least as many points which are first members of pairs as points which are second members. By virtue of the result of problem 83a, the answer to problem 54 is obtained immediately from this,

Inasmuch as the answer to problem 53b can be derived from the answer to problem 54 (compare with the solution to problem 54), the reasoning indicated here also solves problem 53b.

b. The required number G_n of ways is

$$\frac{(2n + 2)(2n + 3) \cdots 3n}{n!} = \frac{1}{2n + 1}\binom{3n}{n}.$$

This result is obtained from the solution to problem 83c in exactly the same way as the solution to problem 54 is obtained from the solution to problem 83a.

c. The required number S_n of ways equals

$$\frac{2n(2n + 1) \cdots (3n - 3)}{(n - 1)!} = \frac{1}{2n - 1}\binom{3n - 3}{n - 1}.$$

To prove this fact one needs only to show that $S_n = G_{n-1}$ (compare with the solution to problem 54).

85a. $\dfrac{1}{m + n}$.

b. Show that the desired probability P_k is independent of k. From this it follows that $P_k = \dfrac{1}{m + n}$.

86. 1/3; 5/9.

87a. 1/5; 1/100.

    b. 1/5; 2/5.

88. 1/7; 91/144.

89. 1/4; 1/20.

90. log 2 (logarithms to the base 10). To prove this fact one must make use of the fact that among the powers of 2 having a given number of digits, exactly one will begin with the figure 1.

91a. Denote by M the number formed by the given sequence of digits. One must show that for any M there are two positive integers n and k such that $10^k M \leq 2^n < 10^k(M + 1)$. Taking logarithms of this inequality, we obtain:

$$\log M + k \leq n \log 2 < \log (M + 1) + k.$$

The problem will be solved if n and k can be found which satisfy the latter inequality. We must therefore show that at least one of the points log 2, 2 log 2, 3 log 2, ... on the real line lies within one of the intervals $[\log M + k, \log (M + 1) + k)$, $k = 1, 2, \cdots$.

    b. log $(1 + 1/M)$ (logarithm to the base 10). Prove that the probability of a randomly selected point n log 2 lying in one of the intervals indicated above equals the length of one of those intervals.

92. Show that the limit in question is

$$\left(1 - \frac{1}{2^2}\right)\left(1 - \frac{1}{3^2}\right)\left(1 - \frac{1}{5^2}\right)\left(1 - \frac{1}{7^2}\right)\left(1 - \frac{1}{11^2}\right)\cdots,$$

where 2, 3, 5, 7, 11, . . . are the prime numbers.

93. Expand $1/(1 - 1/p^2)$ into a geometric series $1 + 1/p^2 + 1/p^4 + 1/p^6 + \cdots$ and use the fundamental theorem of arithmetic. $s = 0.6$ to within 0.1.

94. Show that if $p_1, p_2, \ldots, p_m$ are the first m primes, then

$$\pi(N) \leq m + N - \left[\frac{N}{p_1}\right] - \cdots - \left[\frac{N}{p_m}\right]$$
$$+ \left[\frac{N}{p_1 p_2}\right] + \cdots + (-1)^m\left[\frac{N}{p_1 \cdots p_m}\right].$$

It will be found convenient to throw away the brackets and make an estimate of the maximum error this can entail. Finally, assign a suitable value to m (depending on N).

95. 7/16. The set of all possible outcomes to the experiment is represented here as the set of all points within or on the square $0 \leq x \leq 1, 0 \leq y \leq 1$, where x and y denote the arrival times of the first and second persons respectively and are used as rectangular coordinates.

96. 1/4.

97. The required probability equals

$$\begin{cases} 0 & \text{if} \quad 0 \leq a \leq l/3, \\ \left(3\dfrac{a}{l} - 1\right)^2 & \text{if} \quad l/3 \leq a \leq l/2, \\ 1 - 3\left(1 - \dfrac{a}{l}\right)^2 & \text{if} \quad l/2 \leq a \leq l. \end{cases}$$

98. 1/4. (Compare with problem 96).

99. 1/2. Here all possible outcomes to the experiment are represented by the points of a cube.

100. $1 - (\pi/4)$. (See the hint to the preceding problem.)

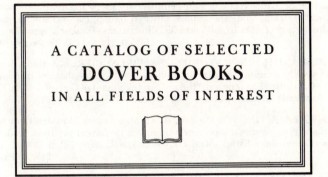

A CATALOG OF SELECTED
DOVER BOOKS
IN ALL FIELDS OF INTEREST

A CATALOG OF SELECTED DOVER
BOOKS IN ALL FIELDS OF INTEREST

DRAWINGS OF REMBRANDT, edited by Seymour Slive. Updated Lippmann, Hofstede de Groot edition, with definitive scholarly apparatus. All portraits, biblical sketches, landscapes, nudes. Oriental figures, classical studies, together with selection of work by followers. 550 illustrations. Total of 630pp. 9⅛ × 12¼.
21485-0, 21486-9 Pa., Two-vol. set $25.00

GHOST AND HORROR STORIES OF AMBROSE BIERCE, Ambrose Bierce. 24 tales vividly imagined, strangely prophetic, and decades ahead of their time in technical skill: "The Damned Thing," "An Inhabitant of Carcosa," "The Eyes of the Panther," "Moxon's Master," and 20 more. 199pp. 5⅜ × 8½. 20767-6 Pa. $3.95

ETHICAL WRITINGS OF MAIMONIDES, Maimonides. Most significant ethical works of great medieval sage, newly translated for utmost precision, readability. Laws Concerning Character Traits, Eight Chapters, more. 192pp. 5⅜ × 8½.
24522-5 Pa. $4.50

THE EXPLORATION OF THE COLORADO RIVER AND ITS CANYONS, J. W. Powell. Full text of Powell's 1,000-mile expedition down the fabled Colorado in 1869. Superb account of terrain, geology, vegetation, Indians, famine, mutiny, treacherous rapids, mighty canyons, during exploration of last unknown part of continental U.S. 400pp. 5⅜ × 8½. 20094-9 Pa. $6.95

HISTORY OF PHILOSOPHY, Julián Marías. Clearest one-volume history on the market. Every major philosopher and dozens of others, to Existentialism and later. 505pp. 5⅜ × 8½. 21739-6 Pa. $8.50

ALL ABOUT LIGHTNING, Martin A. Uman. Highly readable non-technical survey of nature and causes of lightning, thunderstorms, ball lightning, St. Elmo's Fire, much more. Illustrated. 192pp. 5⅜ × 8½. 25237-X Pa. $5.95

SAILING ALONE AROUND THE WORLD, Captain Joshua Slocum. First man to sail around the world, alone, in small boat. One of great feats of seamanship told in delightful manner. 67 illustrations. 294pp. 5⅜ × 8½. 20326-3 Pa. $4.50

LETTERS AND NOTES ON THE MANNERS, CUSTOMS AND CONDITIONS OF THE NORTH AMERICAN INDIANS, George Catlin. Classic account of life among Plains Indians: ceremonies, hunt, warfare, etc. 312 plates. 572pp. of text. 6⅛ × 9¼. 22118-0, 22119-9 Pa. Two-vol. set $15.90

ALASKA: The Harriman Expedition, 1899, John Burroughs, John Muir, et al. Informative, engrossing accounts of two-month, 9,000-mile expedition. Native peoples, wildlife, forests, geography, salmon industry, glaciers, more. Profusely illustrated. 240 black-and-white line drawings. 124 black-and-white photographs. 3 maps. Index. 576pp. 5⅜ × 8½. 25109-8 Pa. $11.95

THE BOOK OF BEASTS: Being a Translation from a Latin Bestiary of the Twelfth Century, T. H. White. Wonderful catalog real and fanciful beasts: manticore, griffin, phoenix, amphivius, jaculus, many more. White's witty erudite commentary on scientific, historical aspects. Fascinating glimpse of medieval mind. Illustrated. 296pp. 5⅜ × 8¼. (Available in U.S. only)   24609-4 Pa. $5.95

FRANK LLOYD WRIGHT: ARCHITECTURE AND NATURE With 160 Illustrations, Donald Hoffmann. Profusely illustrated study of influence of nature—especially prairie—on Wright's designs for Fallingwater, Robie House, Guggenheim Museum, other masterpieces. 96pp. 9¼ × 10¾.   25098-9 Pa. $7.95

FRANK LLOYD WRIGHT'S FALLINGWATER, Donald Hoffmann. Wright's famous waterfall house: planning and construction of organic idea. History of site, owners, Wright's personal involvement. Photographs of various stages of building. Preface by Edgar Kaufmann, Jr. 100 illustrations. 112pp. 9¼ × 10.
                                                                                                23671-4 Pa. $7.95

YEARS WITH FRANK LLOYD WRIGHT: Apprentice to Genius, Edgar Tafel. Insightful memoir by a former apprentice presents a revealing portrait of Wright the man, the inspired teacher, the greatest American architect. 372 black-and-white illustrations. Preface. Index. vi + 228pp. 8¼ × 11.   24801-1 Pa. $9.95

THE STORY OF KING ARTHUR AND HIS KNIGHTS, Howard Pyle. Enchanting version of King Arthur fable has delighted generations with imaginative narratives of exciting adventures and unforgettable illustrations by the author. 41 illustrations. xviii + 313pp. 6⅛ × 9¼.   21445-1 Pa. $5.95

THE GODS OF THE EGYPTIANS, E. A. Wallis Budge. Thorough coverage of numerous gods of ancient Egypt by foremost Egyptologist. Information on evolution of cults, rites and gods; the cult of Osiris; the Book of the Dead and its rites; the sacred animals and birds; Heaven and Hell; and more. 956pp. 6⅛ × 9¼.
                                                                          22055-9, 22056-7 Pa., Two-vol. set $20.00

A THEOLOGICO-POLITICAL TREATISE, Benedict Spinoza. Also contains unfinished *Political Treatise*. Great classic on religious liberty, theory of government on common consent. R. Elwes translation. Total of 421pp. 5⅜ × 8½.
                                                                                                20249-6 Pa. $6.95

INCIDENTS OF TRAVEL IN CENTRAL AMERICA, CHIAPAS, AND YUCATAN, John L. Stephens. Almost single-handed discovery of Maya culture; exploration of ruined cities, monuments, temples; customs of Indians. 115 drawings. 892pp. 5⅜ × 8½.   22404-X, 22405-8 Pa., Two-vol. set $15.90

LOS CAPRICHOS, Francisco Goya. 80 plates of wild, grotesque monsters and caricatures. Prado manuscript included. 183pp. 6⅜ × 9⅜.   22384-1 Pa. $4.95

AUTOBIOGRAPHY: The Story of My Experiments with Truth, Mohandas K. Gandhi. Not hagiography, but Gandhi in his own words. Boyhood, legal studies, purification, the growth of the Satyagraha (nonviolent protest) movement. Critical, inspiring work of the man who freed India. 480pp. 5⅜ × 8½. (Available in U.S. only)
                                                                                                24593-4 Pa. $6.95

ILLUSTRATED DICTIONARY OF HISTORIC ARCHITECTURE, edited by Cyril M. Harris. Extraordinary compendium of clear, concise definitions for over 5,000 important architectural terms complemented by over 2,000 line drawings. Covers full spectrum of architecture from ancient ruins to 20th-century Modernism. Preface. 592pp. 7½ × 9⅝. 24444-X Pa. $14.95

THE NIGHT BEFORE CHRISTMAS, Clement Moore. Full text, and woodcuts from original 1848 book. Also critical, historical material. 19 illustrations. 40pp. 4⅝ × 6. 22797-9 Pa. $2.25

THE LESSON OF JAPANESE ARCHITECTURE: 165 Photographs, Jiro Harada. Memorable gallery of 165 photographs taken in the 1930's of exquisite Japanese homes of the well-to-do and historic buildings. 13 line diagrams. 192pp. 8⅜ × 11¼. 24778-3 Pa. $8.95

THE AUTOBIOGRAPHY OF CHARLES DARWIN AND SELECTED LETTERS, edited by Francis Darwin. The fascinating life of eccentric genius composed of an intimate memoir by Darwin (intended for his children); commentary by his son, Francis; hundreds of fragments from notebooks, journals, papers; and letters to and from Lyell, Hooker, Huxley, Wallace and Henslow. xi + 365pp. 5⅜ × 8. 20479-0 Pa. $5.95

WONDERS OF THE SKY: Observing Rainbows, Comets, Eclipses, the Stars and Other Phenomena, Fred Schaaf. Charming, easy-to-read poetic guide to all manner of celestial events visible to the naked eye. Mock suns, glories, Belt of Venus, more. Illustrated. 299pp. 5¼ × 8¼. 24402-4 Pa. $7.95

BURNHAM'S CELESTIAL HANDBOOK, Robert Burnham, Jr. Thorough guide to the stars beyond our solar system. Exhaustive treatment. Alphabetical by constellation: Andromeda to Cetus in Vol. 1; Chamaeleon to Orion in Vol. 2; and Pavo to Vulpecula in Vol. 3. Hundreds of illustrations. Index in Vol. 3. 2,000pp. 6⅛ × 9¼. 23567-X, 23568-8, 23673-0 Pa., Three-vol. set $36.85

STAR NAMES: Their Lore and Meaning, Richard Hinckley Allen. Fascinating history of names various cultures have given to constellations and literary and folkloristic uses that have been made of stars. Indexes to subjects. Arabic and Greek names. Biblical references. Bibliography. 563pp. 5⅜ × 8½. 21079-0 Pa. $7.95

THIRTY YEARS THAT SHOOK PHYSICS: The Story of Quantum Theory, George Gamow. Lucid, accessible introduction to influential theory of energy and matter. Careful explanations of Dirac's anti-particles, Bohr's model of the atom, much more. 12 plates. Numerous drawings. 240pp. 5⅜ × 8½. 24895-X Pa. $4.95

CHINESE DOMESTIC FURNITURE IN PHOTOGRAPHS AND MEASURED DRAWINGS, Gustav Ecke. A rare volume, now affordably priced for antique collectors, furniture buffs and art historians. Detailed review of styles ranging from early Shang to late Ming. Unabridged republication. 161 black-and-white drawings, photos. Total of 224pp. 8⅜ × 11¼. (Available in U.S. only) 25171-3 Pa. $12.95

VINCENT VAN GOGH: A Biography, Julius Meier-Graefe. Dynamic, penetrating study of artist's life, relationship with brother, Theo, painting techniques, travels, more. Readable, engrossing. 160pp. 5⅜ × 8½. (Available in U.S. only) 25253-1 Pa. $3.95

HOW TO WRITE, Gertrude Stein. Gertrude Stein claimed anyone could understand her unconventional writing—here are clues to help. Fascinating improvisations, language experiments, explanations illuminate Stein's craft and the art of writing. Total of 414pp. 4⅝ × 6⅝. 23144-5 Pa. $5.95

ADVENTURES AT SEA IN THE GREAT AGE OF SAIL: Five Firsthand Narratives, edited by Elliot Snow. Rare true accounts of exploration, whaling, shipwreck, fierce natives, trade, shipboard life, more. 33 illustrations. Introduction. 353pp. 5⅜ × 8½. 25177-2 Pa. $7.95

THE HERBAL OR GENERAL HISTORY OF PLANTS, John Gerard. Classic descriptions of about 2,850 plants—with over 2,700 illustrations—includes Latin and English names, physical descriptions, varieties, time and place of growth, more. 2,706 illustrations. xlv + 1,678pp. 8½ × 12¼. 23147-X Cloth. $75.00

DOROTHY AND THE WIZARD IN OZ, L. Frank Baum. Dorothy and the Wizard visit the center of the Earth, where people are vegetables, glass houses grow and Oz characters reappear. Classic sequel to *Wizard of Oz*. 256pp. 5⅜ × 8. 24714-7 Pa. $4.95

SONGS OF EXPERIENCE: Facsimile Reproduction with 26 Plates in Full Color, William Blake. This facsimile of Blake's original "Illuminated Book" reproduces 26 full-color plates from a rare 1826 edition. Includes "The Tyger," "London," "Holy Thursday," and other immortal poems. 26 color plates. Printed text of poems. 48pp. 5¼ × 7. 24636-1 Pa. $3.50

SONGS OF INNOCENCE, William Blake. The first and most popular of Blake's famous "Illuminated Books," in a facsimile edition reproducing all 31 brightly colored plates. Additional printed text of each poem. 64pp. 5¼ × 7. 22764-2 Pa. $3.50

PRECIOUS STONES, Max Bauer. Classic, thorough study of diamonds, rubies, emeralds, garnets, etc.: physical character, occurrence, properties, use, similar topics. 20 plates, 8 in color. 94 figures. 659pp. 6⅛ × 9¼. 21910-0, 21911-9 Pa., Two-vol. set $14.90

ENCYCLOPEDIA OF VICTORIAN NEEDLEWORK, S. F. A. Caulfeild and Blanche Saward. Full, precise descriptions of stitches, techniques for dozens of needlecrafts—most exhaustive reference of its kind. Over 800 figures. Total of 679pp. 8⅛ × 11. Two volumes. Vol. 1 22800-2 Pa. $10.95
Vol. 2 22801-0 Pa. $10.95

THE MARVELOUS LAND OF OZ, L. Frank Baum. Second Oz book, the Scarecrow and Tin Woodman are back with hero named Tip, Oz magic. 136 illustrations. 287pp. 5⅜ × 8½. 20692-0 Pa. $5.95

WILD FOWL DECOYS, Joel Barber. Basic book on the subject, by foremost authority and collector. Reveals history of decoy making and rigging, place in American culture, different kinds of decoys, how to make them, and how to use them. 140 plates. 156pp. 7⅞ × 10¾. 20011-6 Pa. $7.95

HISTORY OF LACE, Mrs. Bury Palliser. Definitive, profusely illustrated chronicle of lace from earliest times to late 19th century. Laces of Italy, Greece, England, France, Belgium, etc. Landmark of needlework scholarship. 266 illustrations. 672pp. 6⅛ × 9¼. 24742-2 Pa. $14.95

ILLUSTRATED GUIDE TO SHAKER FURNITURE, Robert Meader. All furniture and appurtenances, with much on unknown local styles. 235 photos. 146pp. 9 × 12. 22819-3 Pa. $7.95

WHALE SHIPS AND WHALING: A Pictorial Survey, George Francis Dow. Over 200 vintage engravings, drawings, photographs of barks, brigs, cutters, other vessels. Also harpoons, lances, whaling guns, many other artifacts. Comprehensive text by foremost authority. 207 black-and-white illustrations. 288pp. 6 × 9. 24808-9 Pa. $8.95

THE BERTRAMS, Anthony Trollope. Powerful portrayal of blind self-will and thwarted ambition includes one of Trollope's most heartrending love stories. 497pp. 5⅜ × 8½. 25119-5 Pa. $8.95

ADVENTURES WITH A HAND LENS, Richard Headstrom. Clearly written guide to observing and studying flowers and grasses, fish scales, moth and insect wings, egg cases, buds, feathers, seeds, leaf scars, moss, molds, ferns, common crystals, etc.—all with an ordinary, inexpensive magnifying glass. 209 exact line drawings aid in your discoveries. 220pp. 5⅜ × 8½. 23330-8 Pa. $3.95

RODIN ON ART AND ARTISTS, Auguste Rodin. Great sculptor's candid, wide-ranging comments on meaning of art; great artists; relation of sculpture to poetry, painting, music; philosophy of life, more. 76 superb black-and-white illustrations of Rodin's sculpture, drawings and prints. 119pp. 8⅜ × 11¼. 24487-3 Pa. $6.95

FIFTY CLASSIC FRENCH FILMS, 1912–1982: A Pictorial Record, Anthony Slide. Memorable stills from Grand Illusion, Beauty and the Beast, Hiroshima, Mon Amour, many more. Credits, plot synopses, reviews, etc. 160pp. 8¼ × 11. 25256-6 Pa. $11.95

THE PRINCIPLES OF PSYCHOLOGY, William James. Famous long course complete, unabridged. Stream of thought, time perception, memory, experimental methods; great work decades ahead of its time. 94 figures. 1,391pp. 5⅜ × 8½. 20381-6, 20382-4 Pa., Two-vol. set $19.90

BODIES IN A BOOKSHOP, R. T. Campbell. Challenging mystery of blackmail and murder with ingenious plot and superbly drawn characters. In the best tradition of British suspense fiction. 192pp. 5⅜ × 8½. 24720-1 Pa. $3.95

CALLAS: PORTRAIT OF A PRIMA DONNA, George Jellinek. Renowned commentator on the musical scene chronicles incredible career and life of the most controversial, fascinating, influential operatic personality of our time. 64 black-and-white photographs. 416pp. 5⅜ × 8¼. 25047-4 Pa. $7.95

GEOMETRY, RELATIVITY AND THE FOURTH DIMENSION, Rudolph Rucker. Exposition of fourth dimension, concepts of relativity as Flatland characters continue adventures. Popular, easily followed yet accurate, profound. 141 illustrations. 133pp. 5⅜ × 8½. 23400-2 Pa. $3.50

HOUSEHOLD STORIES BY THE BROTHERS GRIMM, with pictures by Walter Crane. 53 classic stories—Rumpelstiltskin, Rapunzel, Hansel and Gretel, the Fisherman and his Wife, Snow White, Tom Thumb, Sleeping Beauty, Cinderella, and so much more—lavishly illustrated with original 19th century drawings. 114 illustrations. x + 269pp. 5⅜ × 8½. 21080-4 Pa. $4.50

SUNDIALS, Albert Waugh. Far and away the best, most thorough coverage of ideas, mathematics concerned, types, construction, adjusting anywhere. Over 100 illustrations. 230pp. 5⅜ × 8½. 22947-5 Pa. $4.00

PICTURE HISTORY OF THE NORMANDIE: With 190 Illustrations, Frank O. Braynard. Full story of legendary French ocean liner: Art Deco interiors, design innovations, furnishings, celebrities, maiden voyage, tragic fire, much more. Extensive text. 144pp. 8⅞ × 11¾. 25257-4 Pa. $9.95

THE FIRST AMERICAN COOKBOOK: A Facsimile of "American Cookery," 1796, Amelia Simmons. Facsimile of the first American-written cookbook published in the United States contains authentic recipes for colonial favorites—pumpkin pudding, winter squash pudding, spruce beer, Indian slapjacks, and more. Introductory Essay and Glossary of colonial cooking terms. 80pp. 5⅜ × 8½. 24710-4 Pa. $3.50

101 PUZZLES IN THOUGHT AND LOGIC, C. R. Wylie, Jr. Solve murders and robberies, find out which fishermen are liars, how a blind man could possibly identify a color—purely by your own reasoning! 107pp. 5⅜ × 8½. 20367-0 Pa. $2.00

THE BOOK OF WORLD-FAMOUS MUSIC—CLASSICAL, POPULAR AND FOLK, James J. Fuld. Revised and enlarged republication of landmark work in musico-bibliography. Full information about nearly 1,000 songs and compositions including first lines of music and lyrics. New supplement. Index. 800pp. 5⅜ × 8¼. 24857-7 Pa. $14.95

ANTHROPOLOGY AND MODERN LIFE, Franz Boas. Great anthropologist's classic treatise on race and culture. Introduction by Ruth Bunzel. Only inexpensive paperback edition. 255pp. 5⅜ × 8½. 25245-0 Pa. $5.95

THE TALE OF PETER RABBIT, Beatrix Potter. The inimitable Peter's terrifying adventure in Mr. McGregor's garden, with all 27 wonderful, full-color Potter illustrations. 55pp. 4¼ × 5½. (Available in U.S. only) 22827-4 Pa. $1.75

THREE PROPHETIC SCIENCE FICTION NOVELS, H. G. Wells. *When the Sleeper Wakes, A Story of the Days to Come* and *The Time Machine* (full version). 335pp. 5⅜ × 8½. (Available in U.S. only) 20605-X Pa. $5.95

APICIUS COOKERY AND DINING IN IMPERIAL ROME, edited and translated by Joseph Dommers Vehling. Oldest known cookbook in existence offers readers a clear picture of what foods Romans ate, how they prepared them, etc. 49 illustrations. 301pp. 6⅛ × 9¼. 23563-7 Pa. $6.00

SHAKESPEARE LEXICON AND QUOTATION DICTIONARY, Alexander Schmidt. Full definitions, locations, shades of meaning of every word in plays and poems. More than 50,000 exact quotations. 1,485pp. 6½ × 9¼.
22726-X, 22727-8 Pa., Two-vol. set $27.90

THE WORLD'S GREAT SPEECHES, edited by Lewis Copeland and Lawrence W. Lamm. Vast collection of 278 speeches from Greeks to 1970. Powerful and effective models; unique look at history. 842pp. 5⅜ × 8½. 20468-5 Pa. $10.95

THE BLUE FAIRY BOOK, Andrew Lang. The first, most famous collection, with many familiar tales: Little Red Riding Hood, Aladdin and the Wonderful Lamp, Puss in Boots, Sleeping Beauty, Hansel and Gretel, Rumpelstiltskin; 37 in all. 138 illustrations. 390pp. 5⅜ × 8½. 21437-0 Pa. $5.95

THE STORY OF THE CHAMPIONS OF THE ROUND TABLE, Howard Pyle. Sir Launcelot, Sir Tristram and Sir Percival in spirited adventures of love and triumph retold in Pyle's inimitable style. 50 drawings, 31 full-page. xviii + 329pp. 6½ × 9¼. 21883-X Pa. $6.95

AUDUBON AND HIS JOURNALS, Maria Audubon. Unmatched two-volume portrait of the great artist, naturalist and author contains his journals, an excellent biography by his granddaughter, expert annotations by the noted ornithologist, Dr. Elliott Coues, and 37 superb illustrations. Total of 1,200pp. 5⅜ × 8.

Vol. I 25143-8 Pa. $8.95
Vol. II 25144-6 Pa. $8.95

GREAT DINOSAUR HUNTERS AND THEIR DISCOVERIES, Edwin H. Colbert. Fascinating, lavishly illustrated chronicle of dinosaur research, 1820's to 1960. Achievements of Cope, Marsh, Brown, Buckland, Mantell, Huxley, many others. 384pp. 5¼ × 8¼. 24701-5 Pa. $6.95

THE TASTEMAKERS, Russell Lynes. Informal, illustrated social history of American taste 1850's–1950's. First popularized categories Highbrow, Lowbrow, Middlebrow. 129 illustrations. New (1979) afterword. 384pp. 6 × 9.

23993-4 Pa. $6.95

DOUBLE CROSS PURPOSES, Ronald A. Knox. A treasure hunt in the Scottish Highlands, an old map, unidentified corpse, surprise discoveries keep reader guessing in this cleverly intricate tale of financial skullduggery. 2 black-and-white maps. 320pp. 5⅜ × 8½. (Available in U.S. only) 25032-6 Pa. $5.95

AUTHENTIC VICTORIAN DECORATION AND ORNAMENTATION IN FULL COLOR: 46 Plates from "Studies in Design," Christopher Dresser. Superb full-color lithographs reproduced from rare original portfolio of a major Victorian designer. 48pp. 9¼ × 12¼. 25083-0 Pa. $7.95

PRIMITIVE ART, Franz Boas. Remains the best text ever prepared on subject, thoroughly discussing Indian, African, Asian, Australian, and, especially, Northern American primitive art. Over 950 illustrations show ceramics, masks, totem poles, weapons, textiles, paintings, much more. 376pp. 5⅜ × 8. 20025-6 Pa. $6.95

SIDELIGHTS ON RELATIVITY, Albert Einstein. Unabridged republication of two lectures delivered by the great physicist in 1920–21. *Ether and Relativity* and *Geometry and Experience*. Elegant ideas in non-mathematical form, accessible to intelligent layman. vi + 56pp. 5⅜ × 8½. 24511-X Pa. $2.95

THE WIT AND HUMOR OF OSCAR WILDE, edited by Alvin Redman. More than 1,000 ripostes, paradoxes, wisecracks: Work is the curse of the drinking classes, I can resist everything except temptation, etc. 258pp. 5⅜ × 8½. 20602-5 Pa. $3.95

ADVENTURES WITH A MICROSCOPE, Richard Headstrom. 59 adventures with clothing fibers, protozoa, ferns and lichens, roots and leaves, much more. 142 illustrations. 232pp. 5⅜ × 8½. 23471-1 Pa. $3.95

PLANTS OF THE BIBLE, Harold N. Moldenke and Alma L. Moldenke. Standard reference to all 230 plants mentioned in Scriptures. Latin name, biblical reference, uses, modern identity, much more. Unsurpassed encyclopedic resource for scholars, botanists, nature lovers, students of Bible. Bibliography. Indexes. 123 black-and-white illustrations. 384pp. 6 × 9. 25069-5 Pa. $8.95

FAMOUS AMERICAN WOMEN: A Biographical Dictionary from Colonial Times to the Present, Robert McHenry, ed. From Pocahontas to Rosa Parks, 1,035 distinguished American women documented in separate biographical entries. Accurate, up-to-date data, numerous categories, spans 400 years. Indices. 493pp. 6½ × 9¼. 24523-3 Pa. $9.95

THE FABULOUS INTERIORS OF THE GREAT OCEAN LINERS IN HISTORIC PHOTOGRAPHS, William H. Miller, Jr. Some 200 superb photographs capture exquisite interiors of world's great "floating palaces"—1890's to 1980's: *Titanic, Ile de France, Queen Elizabeth, United States, Europa,* more. Approx. 200 black-and-white photographs. Captions. Text. Introduction. 160pp. 8⅜ × 11¾. 24756-2 Pa. $9.95

THE GREAT LUXURY LINERS, 1927–1954: A Photographic Record, William H. Miller, Jr. Nostalgic tribute to heyday of ocean liners. 186 photos of Ile de France, Normandie, Leviathan, Queen Elizabeth, United States, many others. Interior and exterior views. Introduction. Captions. 160pp. 9 × 12. 24056-8 Pa. $9.95

A NATURAL HISTORY OF THE DUCKS, John Charles Phillips. Great landmark of ornithology offers complete detailed coverage of nearly 200 species and subspecies of ducks: gadwall, sheldrake, merganser, pintail, many more. 74 full-color plates, 102 black-and-white. Bibliography. Total of 1,920pp. 8⅜ × 11¼. 25141-1, 25142-X Cloth. Two-vol. set $100.00

THE SEAWEED HANDBOOK: An Illustrated Guide to Seaweeds from North Carolina to Canada, Thomas F. Lee. Concise reference covers 78 species. Scientific and common names, habitat, distribution, more. Finding keys for easy identification. 224pp. 5⅜ × 8½. 25215-9 Pa. $5.95

THE TEN BOOKS OF ARCHITECTURE: The 1755 Leoni Edition, Leon Battista Alberti. Rare classic helped introduce the glories of ancient architecture to the Renaissance. 68 black-and-white plates. 336pp. 8⅜ × 11¼. 25239-6 Pa. $14.95

MISS MACKENZIE, Anthony Trollope. Minor masterpieces by Victorian master unmasks many truths about life in 19th-century England. First inexpensive edition in years. 392pp. 5⅜ × 8½. 25201-9 Pa. $7.95

THE RIME OF THE ANCIENT MARINER, Gustave Doré, Samuel Taylor Coleridge. Dramatic engravings considered by many to be his greatest work. The terrifying space of the open sea, the storms and whirlpools of an unknown ocean, the ice of Antarctica, more—all rendered in a powerful, chilling manner. Full text. 38 plates. 77pp. 9¼ × 12. 22305-1 Pa. $4.95

THE EXPEDITIONS OF ZEBULON MONTGOMERY PIKE, Zebulon Montgomery Pike. Fascinating first-hand accounts (1805–6) of exploration of Mississippi River, Indian wars, capture by Spanish dragoons, much more. 1,088pp. 5⅜ × 8½. 25254-X, 25255-8 Pa. Two-vol. set $23.90

A CONCISE HISTORY OF PHOTOGRAPHY: Third Revised Edition, Helmut Gernsheim. Best one-volume history—camera obscura, photochemistry, daguerreotypes, evolution of cameras, film, more. Also artistic aspects—landscape, portraits, fine art, etc. 281 black-and-white photographs. 26 in color. 176pp. 8⅜ × 11¼. 25128-4 Pa. $12.95

THE DORÉ BIBLE ILLUSTRATIONS, Gustave Doré. 241 detailed plates from the Bible: the Creation scenes, Adam and Eve, Flood, Babylon, battle sequences, life of Jesus, etc. Each plate is accompanied by the verses from the King James version of the Bible. 241pp. 9 × 12. 23004-X Pa. $8.95

HUGGER-MUGGER IN THE LOUVRE, Elliot Paul. Second Homer Evans mystery-comedy. Theft at the Louvre involves sleuth in hilarious, madcap caper. "A knockout."—Books. 336pp. 5⅜ × 8½. 25185-3 Pa. $5.95

FLATLAND, E. A. Abbott. Intriguing and enormously popular science-fiction classic explores the complexities of trying to survive as a two-dimensional being in a three-dimensional world. Amusingly illustrated by the author. 16 illustrations. 103pp. 5⅜ × 8½. 20001-9 Pa. $2.00

THE HISTORY OF THE LEWIS AND CLARK EXPEDITION, Meriwether Lewis and William Clark, edited by Elliott Coues. Classic edition of Lewis and Clark's day-by-day journals that later became the basis for U.S. claims to Oregon and the West. Accurate and invaluable geographical, botanical, biological, meteorological and anthropological material. Total of 1,508pp. 5⅜ × 8½. 21268-8, 21269-6, 21270-X Pa. Three-vol. set $25.50

LANGUAGE, TRUTH AND LOGIC, Alfred J. Ayer. Famous, clear introduction to Vienna, Cambridge schools of Logical Positivism. Role of philosophy, elimination of metaphysics, nature of analysis, etc. 160pp. 5⅜ × 8½. (Available in U.S. and Canada only) 20010-8 Pa. $2.95

MATHEMATICS FOR THE NONMATHEMATICIAN, Morris Kline. Detailed, college-level treatment of mathematics in cultural and historical context, with numerous exercises. For liberal arts students. Preface. Recommended Reading Lists. Tables. Index. Numerous black-and-white figures. xvi + 641pp. 5⅜ × 8½. 24823-2 Pa. $11.95

28 SCIENCE FICTION STORIES, H. G. Wells. Novels, *Star Begotten* and *Men Like Gods*, plus 26 short stories: "Empire of the Ants," "A Story of the Stone Age," "The Stolen Bacillus," "In the Abyss," etc. 915pp. 5⅜ × 8½. (Available in U.S. only) 20265-8 Cloth. $10.95

HANDBOOK OF PICTORIAL SYMBOLS, Rudolph Modley. 3,250 signs and symbols, many systems in full; official or heavy commercial use. Arranged by subject. Most in Pictorial Archive series. 143pp. 8⅛ × 11. 23357-X Pa. $5.95

INCIDENTS OF TRAVEL IN YUCATAN, John L. Stephens. Classic (1843) exploration of jungles of Yucatan, looking for evidences of Maya civilization. Travel adventures, Mexican and Indian culture, etc. Total of 669pp. 5⅜ × 8½. 20926-1, 20927-X Pa., Two-vol. set $9.90

CHRISTMAS CUSTOMS AND TRADITIONS, Clement A. Miles. Origin, evolution, significance of religious, secular practices. Caroling, gifts, yule logs, much more. Full, scholarly yet fascinating; non-sectarian. 400pp. 5⅜ × 8½.
23354-5 Pa. $6.50

THE HUMAN FIGURE IN MOTION, Eadweard Muybridge. More than 4,500 stopped-action photos, in action series, showing undraped men, women, children jumping, lying down, throwing, sitting, wrestling, carrying, etc. 390pp. 7⅞ × 10⅝.
20204-6 Cloth. $19.95

THE MAN WHO WAS THURSDAY, Gilbert Keith Chesterton. Witty, fast-paced novel about a club of anarchists in turn-of-the-century London. Brilliant social, religious, philosophical speculations. 128pp. 5⅜ × 8½.
25121-7 Pa. $3.95

A CEZANNE SKETCHBOOK: Figures, Portraits, Landscapes and Still Lifes, Paul Cezanne. Great artist experiments with tonal effects, light, mass, other qualities in over 100 drawings. A revealing view of developing master painter, precursor of Cubism. 102 black-and-white illustrations. 144pp. 8¾ × 6⅜.
24790-2 Pa. $5.95

AN ENCYCLOPEDIA OF BATTLES: Accounts of Over 1,560 Battles from 1479 B.C. to the Present, David Eggenberger. Presents essential details of every major battle in recorded history, from the first battle of Megiddo in 1479 B.C. to Grenada in 1984. List of Battle Maps. New Appendix covering the years 1967–1984. Index. 99 illustrations. 544pp. 6½ × 9¼.
24913-1 Pa. $14.95

AN ETYMOLOGICAL DICTIONARY OF MODERN ENGLISH, Ernest Weekley. Richest, fullest work, by foremost British lexicographer. Detailed word histories. Inexhaustible. Total of 856pp. 6½ × 9¼.
21873-2, 21874-0 Pa., Two-vol. set $17.00

WEBSTER'S AMERICAN MILITARY BIOGRAPHIES, edited by Robert McHenry. Over 1,000 figures who shaped 3 centuries of American military history. Detailed biographies of Nathan Hale, Douglas MacArthur, Mary Hallaren, others. Chronologies of engagements, more. Introduction. Addenda. 1,033 entries in alphabetical order. xi + 548pp. 6½ × 9¼. (Available in U.S. only)
24758-9 Pa. $11.95

LIFE IN ANCIENT EGYPT, Adolf Erman. Detailed older account, with much not in more recent books: domestic life, religion, magic, medicine, commerce, and whatever else needed for complete picture. Many illustrations. 597pp. 5⅜ × 8½.
22632-8 Pa. $8.50

HISTORIC COSTUME IN PICTURES, Braun & Schneider. Over 1,450 costumed figures shown, covering a wide variety of peoples: kings, emperors, nobles, priests, servants, soldiers, scholars, townsfolk, peasants, merchants, courtiers, cavaliers, and more. 256pp. 8⅜ × 11¼.
23150-X Pa. $7.95

THE NOTEBOOKS OF LEONARDO DA VINCI, edited by J. P. Richter. Extracts from manuscripts reveal great genius; on painting, sculpture, anatomy, sciences, geography, etc. Both Italian and English. 186 ms. pages reproduced, plus 500 additional drawings, including studies for *Last Supper*, *Sforza* monument, etc. 860pp. 7⅞ × 10¾. (Available in U.S. only) 22572-0, 22573-9 Pa., Two-vol. set $25.90

THE ART NOUVEAU STYLE BOOK OF ALPHONSE MUCHA: All 72 Plates from "Documents Decoratifs" in Original Color, Alphonse Mucha. Rare copyright-free design portfolio by high priest of Art Nouveau. Jewelry, wallpaper, stained glass, furniture, figure studies, plant and animal motifs, etc. Only complete one-volume edition. 80pp. 9⅜ × 12¼. 24044-4 Pa. $8.95

ANIMALS: 1,419 COPYRIGHT-FREE ILLUSTRATIONS OF MAMMALS, BIRDS, FISH, INSECTS, ETC., edited by Jim Harter. Clear wood engravings present, in extremely lifelike poses, over 1,000 species of animals. One of the most extensive pictorial sourcebooks of its kind. Captions. Index. 284pp. 9 × 12. 23766-4 Pa. $9.95

OBELISTS FLY HIGH, C. Daly King. Masterpiece of American detective fiction, long out of print, involves murder on a 1935 transcontinental flight—"a very thrilling story"—NY Times. Unabridged and unaltered republication of the edition published by William Collins Sons & Co. Ltd., London, 1935. 288pp. 5⅜ × 8½. (Available in U.S. only) 25036-9 Pa. $4.95

VICTORIAN AND EDWARDIAN FASHION: A Photographic Survey, Alison Gernsheim. First fashion history completely illustrated by contemporary photographs. Full text plus 235 photos, 1840–1914, in which many celebrities appear. 240pp. 6½ × 9¼. 24205-6 Pa. $6.00

THE ART OF THE FRENCH ILLUSTRATED BOOK, 1700–1914, Gordon N. Ray. Over 630 superb book illustrations by Fragonard, Delacroix, Daumier, Doré, Grandville, Manet, Mucha, Steinlen, Toulouse-Lautrec and many others. Preface. Introduction. 633 halftones. Indices of artists, authors & titles, binders and provenances. Appendices. Bibliography. 608pp. 8⅜ × 11¼. 25086-5 Pa. $24.95

THE WONDERFUL WIZARD OF OZ, L. Frank Baum. Facsimile in full color of America's finest children's classic. 143 illustrations by W. W. Denslow. 267pp. 5⅜ × 8½. 20691-2 Pa. $5.95

FRONTIERS OF MODERN PHYSICS: New Perspectives on Cosmology, Relativity, Black Holes and Extraterrestrial Intelligence, Tony Rothman, et al. For the intelligent layman. Subjects include: cosmological models of the universe; black holes; the neutrino; the search for extraterrestrial intelligence. Introduction. 46 black-and-white illustrations. 192pp. 5⅜ × 8½. 24587-X Pa. $6.95

THE FRIENDLY STARS, Martha Evans Martin & Donald Howard Menzel. Classic text marshalls the stars together in an engaging, non-technical survey, presenting them as sources of beauty in night sky. 23 illustrations. Foreword. 2 star charts. Index. 147pp. 5⅜ × 8½. 21099-5 Pa. $3.50

FADS AND FALLACIES IN THE NAME OF SCIENCE, Martin Gardner. Fair, witty appraisal of cranks, quacks, and quackeries of science and pseudoscience: hollow earth, Velikovsky, orgone energy, Dianetics, flying saucers, Bridey Murphy, food and medical fads, etc. Revised, expanded In the Name of Science. "A very able and even-tempered presentation."—The New Yorker. 363pp. 5⅜ × 8. 20394-8 Pa. $5.95

ANCIENT EGYPT: ITS CULTURE AND HISTORY, J. E Manchip White. From pre-dynastics through Ptolemies: society, history, political structure, religion, daily life, literature, cultural heritage. 48 plates. 217pp. 5⅜ × 8½. 22548-8 Pa. $4.95

AMERICAN CLIPPER SHIPS: 1833–1858, Octavius T. Howe & Frederick C. Matthews. Fully-illustrated, encyclopedic review of 352 clipper ships from the period of America's greatest maritime supremacy. Introduction. 109 halftones. 5 black-and-white line illustrations. Index. Total of 928pp. 5⅜ × 8½.
25115-2, 25116-0 Pa., Two-vol. set $17.90

TOWARDS A NEW ARCHITECTURE, Le Corbusier. Pioneering manifesto by great architect, near legendary founder of "International School." Technical and aesthetic theories, views on industry, economics, relation of form to function, "mass-production spirit," much more. Profusely illustrated. Unabridged translation of 13th French edition. Introduction by Frederick Etchells. 320pp. 6⅛ × 9¼. (Available in U.S. only)
25023-7 Pa. $8.95

THE BOOK OF KELLS, edited by Blanche Cirker. Inexpensive collection of 32 full-color, full-page plates from the greatest illuminated manuscript of the Middle Ages, painstakingly reproduced from rare facsimile edition. Publisher's Note. Captions. 32pp. 9⅜ × 12¼.
24345-1 Pa. $4.50

BEST SCIENCE FICTION STORIES OF H. G. WELLS, H. G. Wells. Full novel *The Invisible Man,* plus 17 short stories: "The Crystal Egg," "Aepyornis Island," "The Strange Orchid," etc. 303pp. 5⅜ × 8½. (Available in U.S. only)
21531-8 Pa. $4.95

AMERICAN SAILING SHIPS: Their Plans and History, Charles G. Davis. Photos, construction details of schooners, frigates, clippers, other sailcraft of 18th to early 20th centuries—plus entertaining discourse on design, rigging, nautical lore, much more. 137 black-and-white illustrations. 240pp. 6⅛ × 9¼.
24658-2 Pa. $5.95

ENTERTAINING MATHEMATICAL PUZZLES, Martin Gardner. Selection of author's favorite conundrums involving arithmetic, money, speed, etc., with lively commentary. Complete solutions. 112pp. 5⅜ × 8½.
25211-6 Pa. $2.95

THE WILL TO BELIEVE, HUMAN IMMORTALITY, William James. Two books bound together. Effect of irrational on logical, and arguments for human immortality. 402pp. 5⅜ × 8½.
20291-7 Pa. $7.50

THE HAUNTED MONASTERY and THE CHINESE MAZE MURDERS, Robert Van Gulik. 2 full novels by Van Gulik continue adventures of Judge Dee and his companions. An evil Taoist monastery, seemingly supernatural events; overgrown topiary maze that hides strange crimes. Set in 7th-century China. 27 illustrations. 328pp. 5⅜ × 8½.
23502-5 Pa. $5.00

CELEBRATED CASES OF JUDGE DEE (DEE GOONG AN), translated by Robert Van Gulik. Authentic 18th-century Chinese detective novel; Dee and associates solve three interlocked cases. Led to Van Gulik's own stories with same characters. Extensive introduction. 9 illustrations. 237pp. 5⅜ × 8½.
23337-5 Pa. $4.95

Prices subject to change without notice.

Available at your book dealer or write for free catalog to Dept. GI, Dover Publications, Inc., 31 East 2nd St., Mineola, N.Y. 11501. Dover publishes more than 175 books each year on science, elementary and advanced mathematics, biology, music, art, literary history, social sciences and other areas.